Learning How to Use SPSS with Exercises

The Basics of Social Research

SIXTH EDITION

Earl Babbie
Chapman University

Prepared by

Ted Wagenaar
Miami University

WADSWORTH
CENGAGE Learning

Australia • Brazil • Japan • Korea • Mexico • Singapore • Spain • United Kingdom • United States

WADSWORTH
CENGAGE Learning

ISBN-13: 978-1-285-05306-6
ISBN-10: 1-285-05306-0

Wadsworth
20 Davis Drive
Belmont, CA 94002-3098
USA

Cengage Learning is a leading provider of customized learning solutions with office locations around the globe, including Singapore, the United Kingdom, Australia, Mexico, Brazil, and Japan. Locate your local office at: **www.cengage.com/global**

Cengage Learning products are represented in Canada by Nelson Education, Ltd.

To learn more about Wadsworth, visit **www.cengage.com/wadsworth**

Purchase any of our products at your local college store or at our preferred online store **www.cengagebrain.com**

Printed in the United States of America
1 2 3 4 5 6 7 16 15 14 13 12

Contents

Preface iv

Chapter1 Frequency Distributions 1

Chapter 2 Bivariate Analysis 5
Crosstabulation 5
Correlation 7
Comparing Means 8

Chapter 3 Multivariate Analysis 11
Crosstabulation 11
Correlation 13
Regression Analysis 14

Chapter 4 Data Input and Modification 17
Data Input 17
Modifying Data 20

Chapter 5 Chapter Keyed Exercises 25
Chapter 1 Human Inquiry and Science 25
Chapter 2 Paradigms, Theory, and Research 25
Chapter 3 The Ethics and Politics of Social Research 26
Chapter 4 Research Design 26
Chapter 5 Conceptualization, Operationalization, and Measurement 27
Chapter 6 Indexes, Scales, and Typologies 27
Chapter 7 The Logic of Sampling 29
Chapter 8 Experiments 30
Chapter 9 Survey Research 30
Chapter 10 Qualitative Field Research 31
Chapter 11 Unobtrusive Research 31
Chapter 12 Evaluation Research 32
Chapter 13 Qualitative Data Analysis 32
Chapter 14 Quantitative Data Analysis 32
Chapter 15 Reading and Writing Social Research 33

Appendix General Social Survey 35

Preface

Do you want to learn SPSS? Do you have to learn SPSS? You've come to the right place! This volume gives you easily understood instruction in setting up and analyzing data sets using SPSS. Not to worry if your program says PASW instead of SPSS. It's the same program, but renamed to Predictive Analytics SoftWare starting with version 17. With version 19, the name reverted to SPSS. We'll use the more familiar SPSS acronym in this volume.

We will use an interesting and important data set representing random samples of Americans– the General Social Survey. We will start with data analysis because that is the most interesting aspect of SPSS. But if you need to start with data input, head straight to Chapter 4 and then come back to Chapter 1 to learn how to analyze your data set. You will find SPSS easy to master. We address only the basics here, but with a bit of practice you will be able to easily figure out the more advanced topics (assuming you have had instruction in statistics). Chapter 5 contains exercises keyed to Earl Babbie's *The Basics of Social Research,* 6th edition. Most of these exercises can also be used with any Cengage social research methods or statistics text.

The instructions and examples in this volume will work with any version of SPSS. We will use the full samples of the General Social Survey for 2000 and 2010 for examples. If you are using student versions of the program from older versions of SPSS, your instructor will give you a data set that conforms to the limitations of that program–1,500 cases and 50 variables. Be sure to review the Appendix, where the codebook for the General Social Survey is located.

So, be prepared to become a more informed, confident social researcher! Please drop me an email if you have any comments or suggestions for the next version.

Special thanks to Andrea Parks for her Word wizardry skills.

Theodore C. Wagenaar
Miami University
Department of Sociology and Gerontology
Oxford, OH 45056
wagenatc@muohio.edu

Chapter 1

Frequency Distributions

The first task social researchers usually complete after compiling a data set is a frequency distribution. It's also the most common exercise assigned to methods students. If you are looking for advice on setting up an SPSS data set from some data you have collected, we'll cover that in Chapter 4 because not every student is asked to complete that task. Also, we want to start with the fun stuff. You will find that doing data setup and modification is easier after you've actually used SPSS in simpler tasks such as frequency distributions.

Social scientists perform frequency distributions for several reasons. One is to make sure that the coding was done correctly. If we see a code 7 in a frequency distribution for our gender variable, for example, the code is incorrect because we used only codes 1 (for "male" and 2 (for "female"). Using the associated identification number, we could go back to the original survey and might find that someone misread a 7 for what should have been a 1. We can fix the error in the data set before proceeding. Social scientists also perform frequency distributions to get a feel for their data. In your survey of students regarding cheating, for example, you may wish to see the range of responses to your question, "How often have you engaged in academic misconduct?" to see if your initial thoughts about the range of responses are in line with the data. If there is minimal variation in the responses – say all the responses are 1 or 2 on a 12-point scale – it may be difficult to then look for gender or school class level differences. Finally, social scientists look at frequency distributions in case they wish to collapse the responses into fewer categories to make crosstabulation feasible. For example, a frequency distribution on the academic misconduct item might show that one-third of the responses fell between a 3 and 4, another third between 5 and 9, and another third between 10 and 12. We could then use the ***Recode*** command in SPSS to create a collapsed version of the variable, which we could subsequently crosstabulate with a variable such as gender.

Let's get started with data analysis. Load SPSS by accessing ***Programs*** from the ***Start*** menu and clicking SPSS. Or just click the SPSS icon on the desktop if one is there. First we need to open a data set. All exercises in this volume will use the *gss00_10allcases.sav* data file supplied by your instructor. This data set includes all the cases from both the 2000 and 2010 General Social Survey on 116 variables. If you are using a student version of SPSS, your instructor will supply you with another General Social Survey data set containing only 1,500 cases and 50 variables (the limitations imposed by student versions). The variables used as examples in this volume are generally available in both the complete data set and the limited data sets. The codebook for all the data sets is included in the Appendix. Be sure to read the actual questions used instead of relying on the short labels in the SPSS data file because sometimes the labels do not fully capture the meaning of the item.

Figure 1.1 Opening Dialog Box

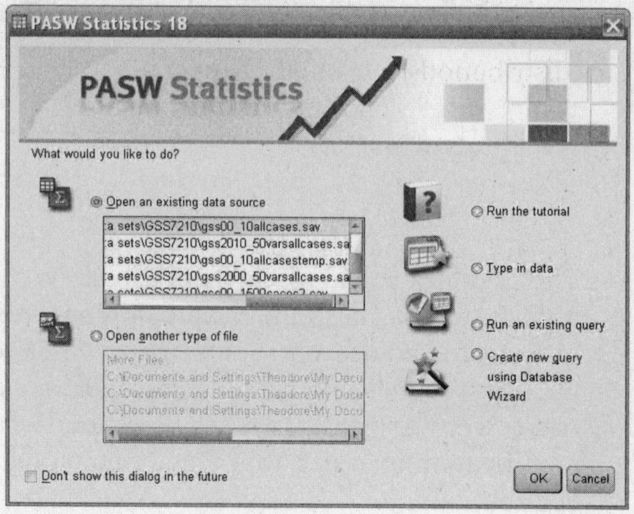

The first screen you will see after opening SPSS is a dialog box asking you what you would like to do (see Figure 1.1). We'll use the default, which is "open an existing data set." Scroll down until you see the *gss00_10allcases.sav* data set and double click it. This action will open the data set. If you do not see this data set among the options, close this dialog box, click ***File*** on the top left of the screen, and then click ***Open*** and ***Data*** and follow your instructor's suggestions for locating the appropriate data set.

Now we're ready to request a frequency distribution. Click ***Analyze*** on the top of the screen, then ***Descriptive Statistics***, then ***Frequencies*** (see Figure 1.2). All the variable names will still be listed along the left of this dialog box. If instead you see the labels for the variables, first change the options to show variable names (unless, of course, you prefer to see the labels). Do this by clicking ***Edit*** at the top, and then ***Options***. Click the ***General*** tab if it is not visible, and under ***Variable Lists*** click ***Display Names*** and ***Alphabetical***. Close the dialog box by clicking ***OK***. Then follow the instructions above for getting the ***Frequencies*** command. Scroll down the list of variables on the left until you get to ATTEND (frequency of attendance to religious services), click on it once to highlight it, and then click on the right arrow in the middle of the dialog box. You will see variable appear on the right in the ***Variables*** box (see Figure 1.3). Your instructor may give you additional instructions for selecting some statistics by clicking the ***Statistics*** box in the dialog box. Then click ***OK***. The results will automatically appear in a separate output window and will look like Figure 1.4.

Figure 1.2 Frequencies Command

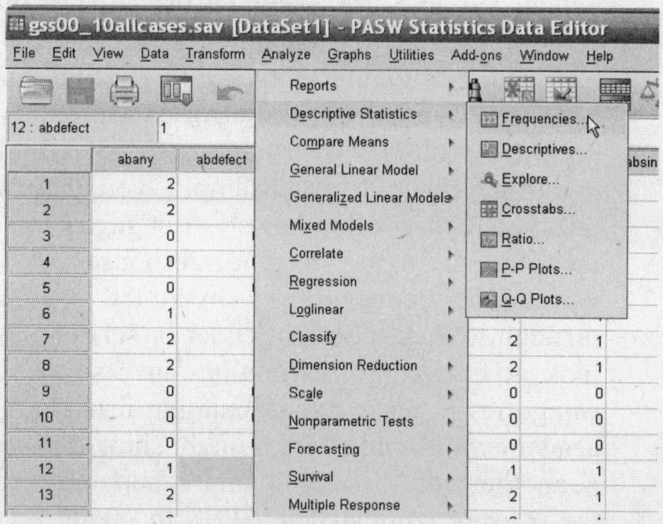

Figure 1.3 Frequencies Dialog Box

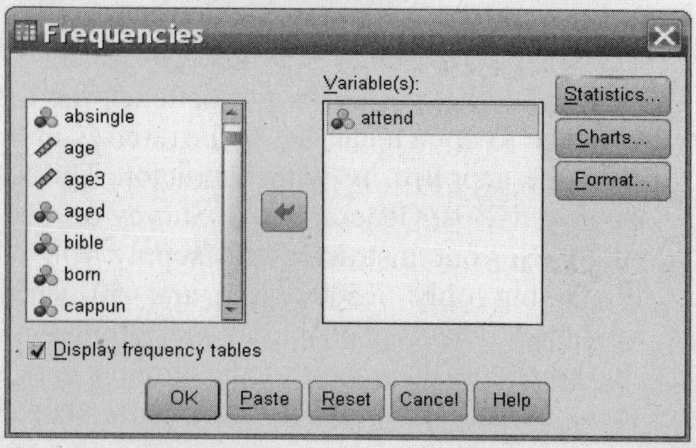

We sometimes forget the details about our variables while trying to decide which variables to select for frequency distributions. If that happens, simply right click a variable in the variables list in the *Frequencies* dialog box and then click ***Variable Information***. Doing so will give you the variable name, the variable description, and the value labels. Another tip: if you accidentally send a variable from the variable list on the left to the window of variables on the right for frequency distributions, no worries. Just highlight the variable you wish to send back to the variable list on the left and then click the left arrow in the middle of the dialog box (it changes direction based on the location of a highlighted variable: variable list on the left or selection window variables on the right). Or you can press the ***Reset*** button and all the variables in the selection window will go back to the variable list window. This reset feature applies to all SPSS dialog boxes.

Figure 1.4 Frequencies Output

Statistics

attend HOW OFTEN R ATTENDS RELIGIOUS SERVICES

N	Missing	88

attend HOW OFTEN R ATTENDS RELIGIOUS SERVICES

		Frequency	Percent	Valid Percent	Cumulative Percent
Valid	1 LT ONCE A YEAR	1415	29.1	29.6	29.6
	2 ONCE A YR THRU SEV TIMES A YEAR	1201	24.7	25.2	54.8
	3 ONCE A MNTH THRU ALMST WEEKLY	954	19.6	20.0	74.8
	4 EVERY WEEK OR MORE	1203	24.7	25.2	100.0
	Total	4773	98.2	100.0	
Missing	9 DK,NA	88	1.8		
Total		4861	100.0		

The output shows that the number of respondents with valid data is 4,861 (88 cases have missing data on this item), and then gives the breakdown for each category of attendance at religious services. We see that 1,415 respondents (29.6%) attend religious services less than once a year, 1,201 respondents (25.2%) attend once a year through several times a year, 954 respondents (20.0%) attend almost once a month through almost weekly, and 1,203 respondents (25.2%) attend every week or more often. The percentage figures reflect the "valid percent" column. When a variable has missing data for some cases, the "percent" column shows the actual percent in each response category including a percentage value for the "missing" category, while the "valid percent" column shows the percent in each response category excluding the "missing" cases. We usually use the "valid percent" category because we wish to examine how the responses are distributed only among those with non-missing data.

Figure 1.5 Sample Bar Chart

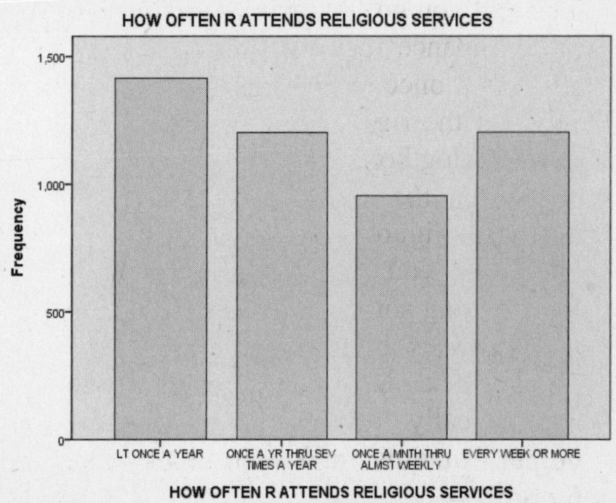

You can also use the ***Frequencies*** command to request various types of charts. While in the ***Frequencies*** dialog box, simply click on the ***Charts*** tab at the bottom of the dialog box and indicate which type of chart you desire. A *bar chart* visually shows the relative frequency of each response category, separated by space, and includes short labels underneath each bar. Figure 1.5 shows a bar chart for religious attendance. A *pie chart* shows how much of a pie-shaped chart each response category reflects. A *histogram* is similar to a bar chart, but has no space between the response categories displayed, and only reports the numerical values of the various responses instead of the labels associated with those values. These are the only three graphical displays available within the ***Frequencies*** command. Note that you can graph either frequencies or percentages in the bar chart or pie chart. More sophisticated charts, as well as more options for the three types available within ***Frequencies***, can be explored by clicking ***Graphs*** at the top of the screen.

Some variables have too many values to allow easy analysis with the ***Frequencies*** command. For example, if you recorded the actual incomes of 1,000 people, you could get 1,000 different values for income in a frequency distribution. The distribution would take many pages to print and may not provide you with the data analysis results you desire. In such cases, it is better to simply request descriptive statisticsBsummary statistics without the actual frequency distribution. To do this in SPSS, click ***Analyze***, then click ***Descriptive Statistics***, and then ***Descriptives***. The procedure for selecting variables from the variable list on the left and moving them to the selection window on the right is the same as in ***Frequencies***. Then click ***OK***. For example, performing this analysis for **AGE** in the General Social Survey data set shows 4,850 valid cases with a minimum age of 18 and a maximum age of 89. The mean age is 46.84 and the standard deviation is 17.52. You can use the ***Options*** button before requesting the descriptive statistics to refine your choice of which statistics to present and how to present them.

Chapter 2

Bivariate Analysis

Bivariate analysis is where the real action is. It helps address one of the prerequisites for establishing causality: does a relationship exist between an independent variable and a dependent variable? The other two are establishing time order and ruling out other possible causes. We will address the latter in the next chapter on multivariate analyses. In this chapter, we will consider several basic types of bivariate analysis techniques available in SPSS: crosstabulation, correlation, and comparison of means.

Crosstabulation

Crosstabulation is simply constructing a table to show the relationship between an independent variable and a dependent variable by using percentage comparisons. Recall the advice from your methods course for constructing tables: 1) put the independent variable on top of the columns, 2) put the dependent variable on the side of the rows, 3) calculate the percentages going down, and 4) read across the columns of the independent variable in terms of the dependent variable. Percentage down means that the percentages *within* one column total 100 percent. Reading across means that you read the percentage value within one column for a particular category of the dependent variable and compare the percentages within the remaining columns on the same category of the dependent variable. Sounds complicated, but it's really quite simple. Let's do an example. Recall that we will use the *gss00_10allcases.sav* data file supplied by your instructor (you may use a different data set if you are using a student version of SPSS). Recall too that the codebook is ported in the Appendix.

Figure 2.1 Crosstab Command

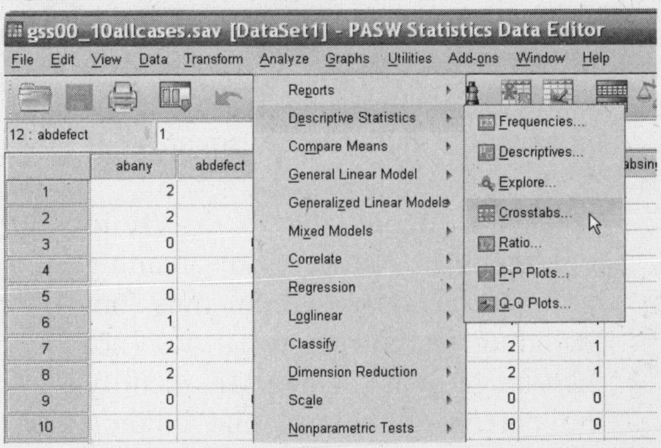

Let's see if there has been a change between 2000 and 2010 regarding support for this item: "Most men are better suited emotionally for politics than are most women." Open the data set as previously described. Then click ***Analyze*** on the top of the screen, then ***Descriptive Statistics***, then ***Crosstabs*** (see Figure 2.1). All the variable names will be listed along the left of this dialog box and you can follow the procedure mentioned before for highlighting a variable on the left and then clicking on the arrow box in the middle to move it to the right (see Figure 2.2). Except now we have three boxes on the right to choose from to place our selected variable. The top box is for the row variable – our dependent variable. The second box is for the column variable – our independent variable. The third box, well, we'll wait until the next chapter to deal with that box B it has to do with multivariate analysis. So, highlight **FEPOL** on the left and move it to the top box, the row variable. Then highlight **YEAR** on the left and move it to the middle box, the column variable. Click the ***Statistics*** box and you will notice a choice of many statistics. For now, just click ***Chi Square***. After closing the ***Statistics*** dialog box, open the ***Cells*** box and click ***Column*** under ***Percentages*** (ignore the other choices for now). Click ***Continue*** and then click ***OK***.

Figure 2.2 Crosstab Dialog Box

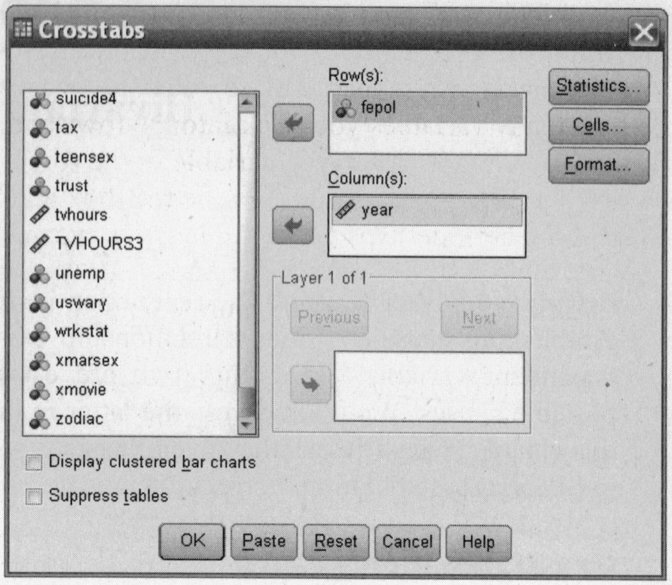

No doubt we will see a shift towards greater acceptance of women in politics in the decade 2000-2010, a possible result of the women's movement. Let's examine the table (see Figure 2.3). To read a table, read the percentage of respondents within the first column that fall within the first row. The results show that 23.3 percent of respondents in 2000 indicated that they agreed that "Most men are better suited emotionally for politics than are most women." Then read the percentage of respondents within the second column that also fall within the first row. In this case, 21.2 percent of respondents in 2010 indicated that they agreed with the item. The results show a slight decrease in agreement with this item over the decade. However, the chi square shows that the change is not statistically significant (your instructor will review statistical significance and how to interpret the values of chi square).

I picked a simple table to analyze to highlight the process of reading tables. It has only two columns (not including totals) and two rows (not including totals). Once we identify the percentage within one column that falls within one row, we can skip analyzing the second row because it will be the opposite of the first row. If a table has more than two rows and two columns, we will need to examine enough of the cells in order to determine if a relationship exists. The techniques used in a simple table also work for a more complex table.

Figure 2.3 Crosstab Output

fepol WOMEN NOT SUITED FOR POLITICS * year GSS YEAR FOR THIS RESPONDENT Crosstabulation

			year GSS YEAR FOR THIS RESPONDENT		
			2000	2010	Total
fepol WOMEN NOT SUITED FOR POLITICS	1 AGREE	Count	403	287	690
		% within year GSS YEAR FOR THIS RESPONDENT	23.3%	21.2%	22.4%
	2 DISAGREE	Count	1327	1064	2391
		% within year GSS YEAR FOR THIS RESPONDENT	76.7%	78.8%	77.6%
Total		Count	1730	1351	3081
		% within year GSS YEAR FOR THIS RESPONDENT	100.0%	100.0%	100.0%

Three cautionary notes. First, it is not always immediately obvious if a variable is independent or dependent. For example, the number of hours someone watches television could be a dependent variable caused by, say, age. But the number of hours of TV watching could itself be an independent variable that potentially influences something like voting preferences or views about gender roles. As noted in your methods course, you must make such choices before you engage in data analysis with SPSS. Second, be careful how many variables you click into the rows and column boxes in SPSS because SPSS will construct a table for every variable in the row box with every variable in the column box. So, if you put five variables in the row box and five variables in the column box, you will get 25 tables, perhaps far more than you wanted. Consider carefully how many tables you want and don't be shy about running the ***Crosstabs*** procedure multiple times with fewer variables in the row and/or column dialog boxes to get the exact tables you need.

Third, I said to request column percentages to facilitate your analysis and to keep things simple. But there may be occasions when you wish to request both column and row percentages. For example, if you are examining the relationship between two items in a scale that measure the same concept, neither is really "independent" or "dependent." Or perhaps you are examining the relationship between self esteem and depression – it's not obvious which is independent and which is dependent. It makes sense in such situations to request both row and column percentages so you can more thoroughly examine the distributions across various combinations of responses in order to complete the analyses you need to answer your research questions.

Correlation

Correlation analysis addresses the same goal as do tables: does a relationship exist between the independent and dependent variables? Correlation analysis helps answer this question without tables, although we often calculate correlations for tables as well. In tables, the correlations statistics summarize the overall relationships between two variables. Two popular correlation coefficients are the Pearson correlation coefficient (used with interval and ratio level variables) and Kendall's tau (used with ordinal level variables or when one variable is ordinal and the other commonly interval or ratio. For example, it is commonly assumed that age and frequency of sex are inversely related. That is, the older people are, the less sex they have. Let's calculate that correlation coefficient.

Figure 2.4 Bivariate Correlation Command

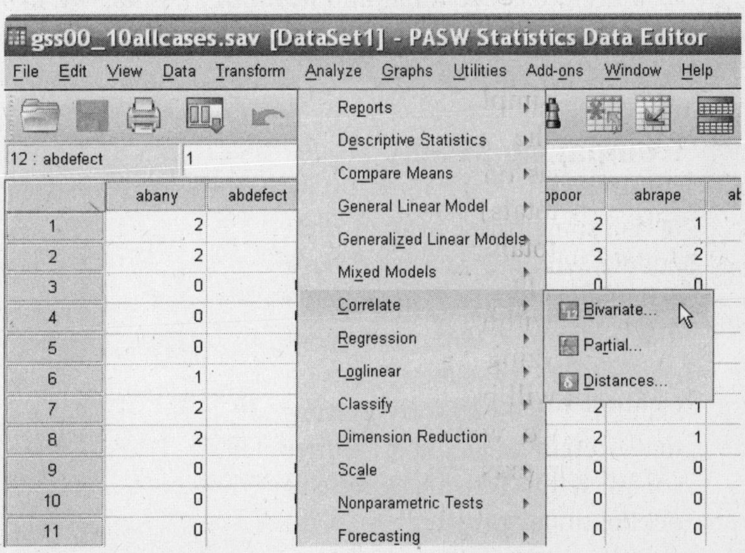

To perform a correlation analysis in SPSS, click ***Analyze*** on the top of the screen, then ***Correlate***, and then ***Bivariate*** (see Figure 2.4). All the variable names will be listed along the left of this dialog box and you can follow the procedure mentioned before for highlighting a variable on the left and then clicking on the arrow box in the middle to move it to the right. Follow this step to place **SEXFREQ** and **AGE** in the selection window on the right (see Figure 2.5). Because **SEXFREQ** is constructed as an ordinal level variable, uncheck the default ***Pearson*** under ***Correlation Coefficients*** and check ***Kendall's tau-b***. You can request means and standard deviations for interval or ratio level variables by clicking the ***Options*** box and clicking ***Means and Standard Deviations*** (such statistics are not very useful for ordinal level variables). Then click ***OK*** to calculate the correlation coefficient. Your results should resemble Figure 2.6.

Figure 2.5 Bivariate Correlations Dialog Box

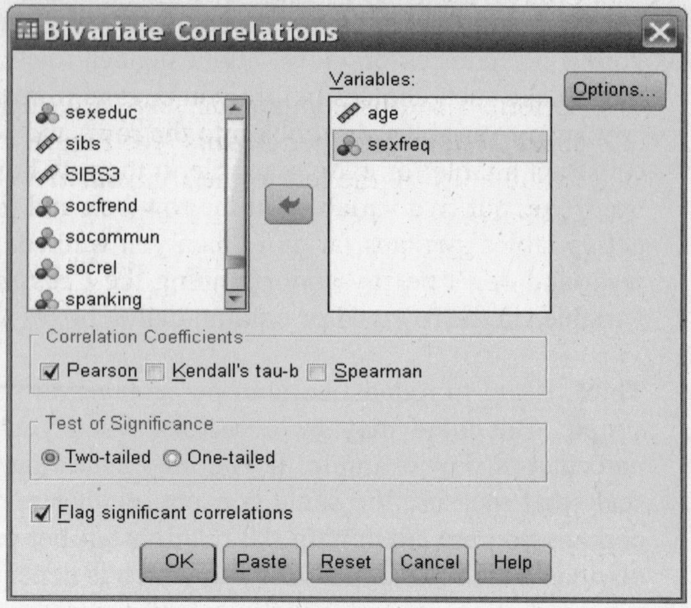

Figure 2.6 Bivariate Correlations Output

Correlations

			age AGE OF RESPONDENT	sexfreq FREQUENCY OF SEX DURING LAST YEAR
Kendall's tau_b	age AGE OF RESPONDENT	Correlation Coefficient	1.000	-.329**
		Sig. (2-tailed)	.	.000
		N	4850	4010
	sexfreq FREQUENCY OF SEX DURING LAST YEAR	Correlation Coefficient	-.329**	1.000
		Sig. (2-tailed)	.000	.
		N	4010	4016

**. Correlation is significant at the 0.01 level (2-tailed).

The resulting correlation coefficient is -.33. The results indicate a modest negative correlation between age and frequency of sex. Because our data are cross-sectional and not longitudinal, we cannot conclude that as people age, they have less sex. At best, we can say only that older people have less sex than do younger people, to a modest extent.

Comparing Means

It is sometimes argued that people who are rewarded by a social system tend to support that system. This statement might be translated into a testable hypothesis by stating that Republicans have higher social status (as measured by occupational prestige) than do members of other political parties. We can examine this relationship in SPSS by using

Figure 2.7 Compare Means Command

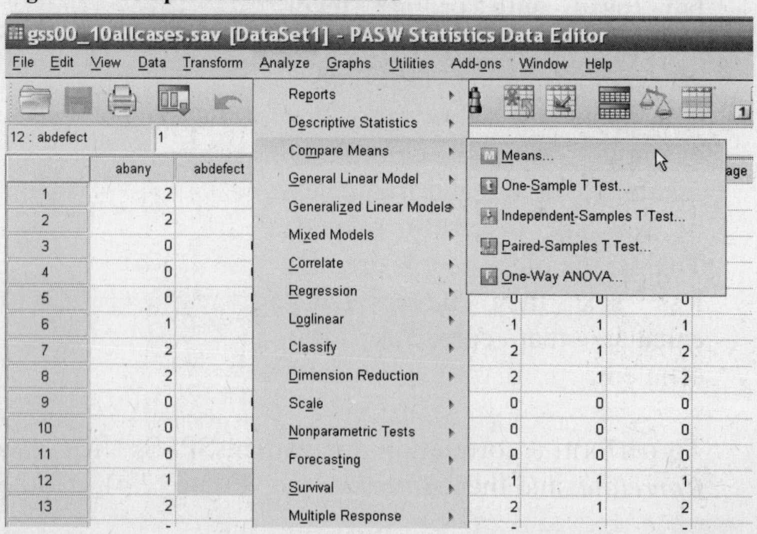

the ***Compare Means*** procedure, which is used when one variable is interval or ratio level and the other is nominal level.

To perform a means comparison in SPSS, click ***Analyze*** on the top of the screen, then ***Compare Means***, and then ***Means*** (see Figure 2.7). Move **SEI** (the prestige variable) from the variable list on the left to the ***Dependent List*** box on the right, and move **PARTYID** to the ***Independent List*** box (see Figure 2.8). Your instructor may advise you to make particular selections in the ***Options*** box. Click ***OK.*** Your results should resemble Figure 2.9.

Figure 2.8 Compare Means Dialog Box

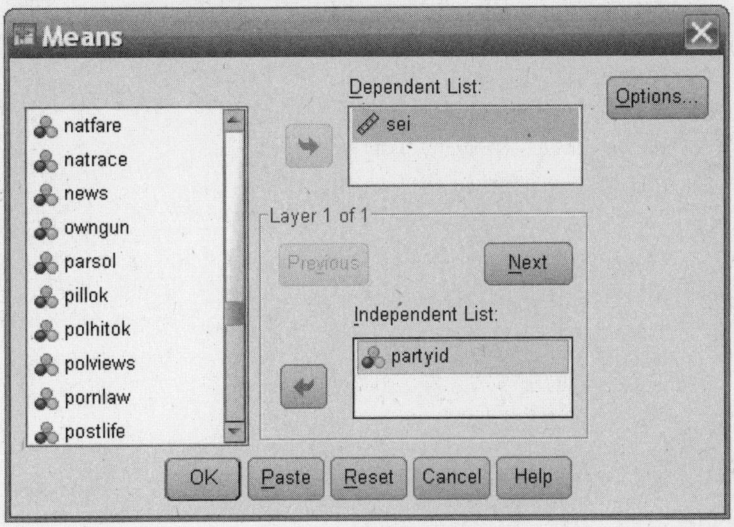

The results show that Republicans do have higher occupational prestige (mean score of 52.7) than Democrats (48.1) or Independents (47.5). Those belonging to another party have a mean occupational prestige score of 54.6 (but note that only 90 respondents are in this category). The hypothesis is supported.

Figure 2.9 Compare Means Output

Case Processing Summary

	Cases					
	Included		Excluded		Total	
	N	Percent	N	Percent	N	Percent
sei RESPONDENT SOCIOECONOMIC INDEX * partyid POLITICAL PARTY AFFILIATION	4505	92.7%	356	7.3%	4861	100.0%

Report

sei RESPONDENT SOCIOECONOMIC INDEX

partyid POLITICAL PAR...	Mean	N	Std. Deviation
1 DEMOCRAT	48.125	1506	19.0503
2 INDEPENDENT	47.481	1829	19.1505
3 REPUBLICAN	52.721	1080	19.2713
4 OTHER PARTY	54.573	90	21.1233
Total	49.094	4505	19.3138

Chapter 3

Multivariate Analysis

Recall that one of the prerequisites for establishing causality is to rule out other possible causes. Multivariate analysis helps address this issue. Multivariate analysis allows several types of investigations. One is to look at a bivariate relationship of interest and see what happens to it in the face of controls for one or more control variables. For example, we may wish to see if a relationship we have previously found between age and religiosity holds up when we add a control for social class. Second, multivariate analysis allows an inspection of the unique impact of each of several independent variables on one dependent variable. Unique impact means the impact of an independent variable while simultaneously controlling for all the others. For example, what is the unique impact of high school grades, SAT scores, and social class on college attendance? Third, we can use multivariate analysis to inspect the interdependency among items. For example, what patterns seem to emerge in a multivariate analysis of several items in a support for abortion scale? The most basic forms of multivariate analysis include crosstabulation, correlation, and multiple regression.

Crosstabulation

This one is easy to do. You already learned how to perform crosstabulations. To make a crosstabulation analysis multivariate, simply add a control variable. Let's look at the relationship between age and agreement or disagreement with this statement: "Do you think that a person has the right to end his or her life if this person is tired of living and ready to die?"

Figure 3.1 Crosstab Dialog Box for Multivariate Analysis

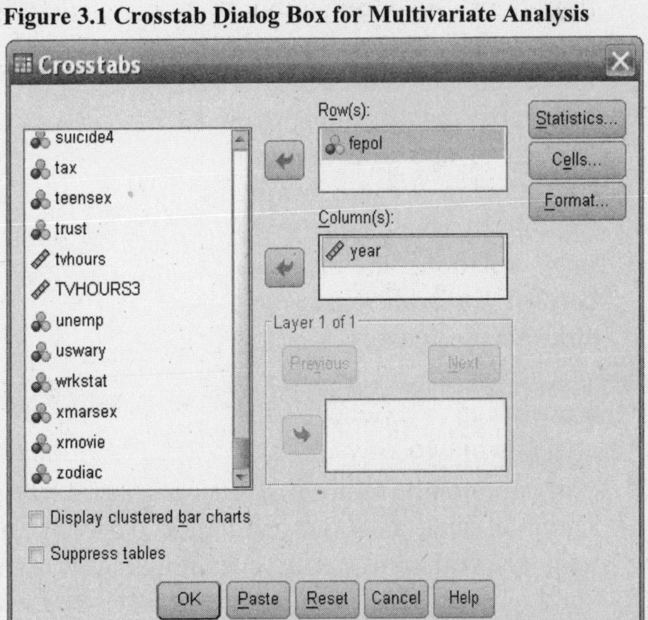

To perform a multivariate cross-tabulation click ***Analyze*** on the top of the screen, then ***Descriptive Statistics***, then ***Crosstabs*** (look back to Figure 2.1). Move **SUICIDE4** to the row variable selection box and move **AGE3** (raw age collapsed into three categories) to the column variable selection box. So far, this looks just like what we did earlier to perform a bivariate analysis. But now move **SEX** to the ***Layer***

selection box to indicate our control variable (see Figure 3.1). As before, click the *Cells* box and select column percentages.

Figure 3.2 Bivariate Table

suicide4 SUICIDE IF TIRED OF LIVING * age3 AGE IN THIRDS Crosstabulation

			age3 AGE IN THIRDS			
			1 18 THRU 36	2 37 THRU 53	3 54 THRU 89	Total
suicide4 SUICIDE IF TIRED OF LIVING	1 YES	Count	199	181	184	564
		% within age3 AGE IN THIRDS	19.4%	16.2%	17.7%	17.7%
	2 NO	Count	828	933	857	2618
		% within age3 AGE IN THIRDS	80.6%	83.8%	82.3%	82.3%
Total		Count	1027	1114	1041	3182
		% within age3 AGE IN THIRDS	100.0%	100.0%	100.0%	100.0%

In order to make comparisons with **SUICIDE4**), let's also do a bivariate crosstab (see Figure 3.2 for the results). That table shows modest differences: 19.4 percent of 18-36 year olds agree that someone should have the right to commit suicide if that person is tired of living compared to 16.2 percent of 37-52 year olds and 17.7 percent of 53-89 year olds. Hence, young people are slightly more likely to support suicide than older people. The second table we get is a repeat of the first table, but broken down by **SEX** (see Figure 3.3). Now we see that younger men are slightly more likely to agree with the statement than middle age or older men (22.4% versus 18.2% and 20.8%). We also see that younger women are somewhat more likely to agree with the statement than middle age or older women (17.1% versus 14.6% and 15.6%). Looking at the totals column on the right, we also see that men overall are more likely to agree with the statement than are women overall (20.4% versus 15.7%).

The small age difference in suicide support shown in the bivariate table is repeated in the multivariate table, where we see similar age differences among men and women. Your instructor will review the various outcomes possible as we shift from bivariate to multivariate analysis. Important to remember, however, is that we are still interested in our original bivariate relationship (in this case, age and support for suicide). We are simply looking at what happens to that relationship as we control for additional variables (in this case, sex).

Figure 3.3 Trivariate Table

suicide4 SUICIDE IF TIRED OF LIVING * age3 AGE IN THIRDS * sex RESPONDENTS SEX Crosstabulation

				age3 AGE IN THIRDS			
sex RESPONDENTS SEX				1 18 THRU 36	2 37 THRU 53	3 54 THRU 89	Total
1 MALE	suicide4 SUICIDE IF TIRED OF LIVING	1 YES	Count	99	93	85	277
			% within age3 AGE IN THIRDS	22.4%	18.2%	20.8%	20.4%
		2 NO	Count	342	417	323	1082
			% within age3 AGE IN THIRDS	77.6%	81.8%	79.2%	79.6%
	Total		Count	441	510	408	1359
			% within age3 AGE IN THIRDS	100.0%	100.0%	100.0%	100.0%
2 FEMALE	suicide4 SUICIDE IF TIRED OF LIVING	1 YES	Count	100	88	99	287
			% within age3 AGE IN THIRDS	17.1%	14.6%	15.6%	15.7%
		2 NO	Count	486	516	534	1536
			% within age3 AGE IN THIRDS	82.9%	85.4%	84.4%	84.3%
	Total		Count	586	604	633	1823
			% within age3 AGE IN THIRDS	100.0%	100.0%	100.0%	100.0%

You can extend this analysis in two ways. First, you can simply remove **SEX** as a control variable in ***Layer 1*** and replace it with another variable, such as **RACE**, and then perform an analysis like the one above for **SEX**. This analysis will allow you to examine how race affects the relationship between age and suicide support. Second, leaving **SEX** in ***Layer 1***, you could click the ***Next*** box in the ***Layer*** selection window and place **RACE** in ***Layer 2.*** Now you will get the same tables you obtained before — age with suicide support controlling for sex, but with each of those control tables further broken down by race. This analysis would allow you to examine the impact of race on the sex differences in the bivariate relationship between age and suicide support.

Correlation

The same logic applies with correlations, although correlations may be easier to interpret because you do not end up with numerous tables with percentages for each category of independent, dependent, and control variables. Let's examine the correlation between age and TV watching and then control for social class (as measured by occupational prestige). Because social class and age may both influence television watching, let's look at how social class affects the bivariate correlation between age and television watching.

Figure 3.4 Partial Correlation Command

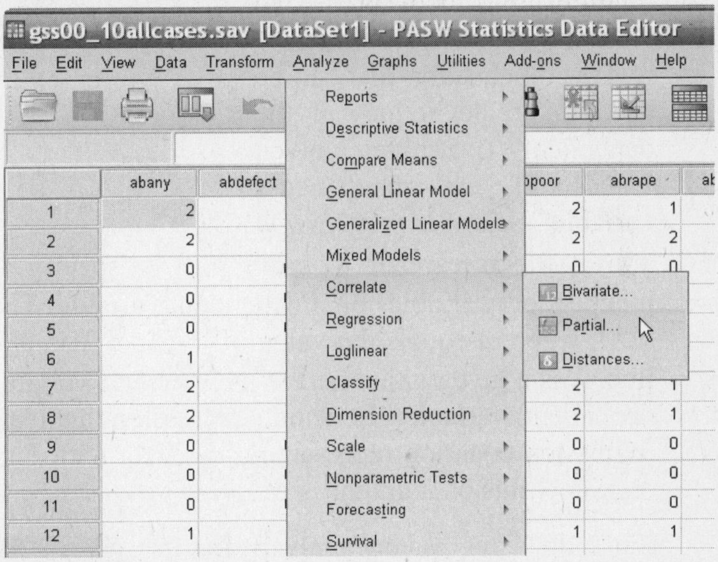

First calculate the bivariate correlation between **AGE** and **TVHOURS** so that we have a base correlation as a reference point. As before, click ***Analyze*** on the top of the screen, then ***Correlate***, and then ***Bivariate*** (look back to Figure 2.4). Move **AGE** and **TVHOURS** to the ***Variables*** selection window and click ***OK***. The resulting correlation is .13, a low but positive correlation. To a minimal extent, older people watch more television.

Figure 3.5 Partial Correlation Dialog Box

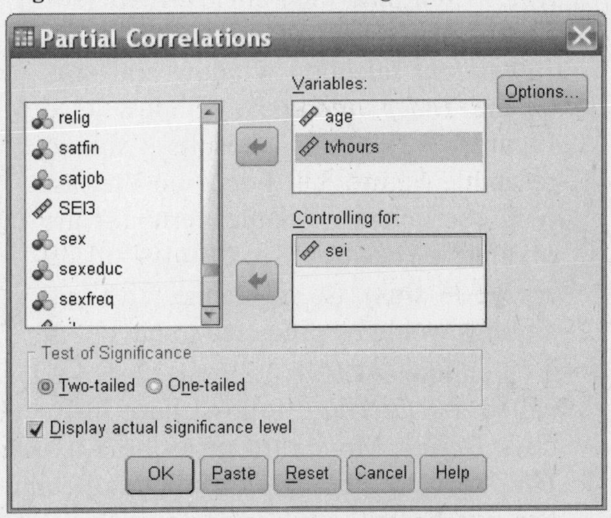

Now perform a multivariate analysis by again requesting the correlation between **AGE** and **TVHOURS**, but now controlling for **SEI** (social class). Click ***Analyze***, click ***Correlate***, and then click ***Partial*** (see Figure 3.4). Next, move **AGE** and **TVHOURS** to the ***Variables*** selection window and move **SEI** to the ***Controlling for*** selection window (see Figure 3.5). The resulting correlation between age and television watching, controlling for social class, is .16. The original low correlation again occurs, but at least now we know that social class does not affect this bivariate relationship. This helps us to be a little more sure about the original relationship between age and television watching.

Regression Analysis

Figure 3.6 Linear Regression Command

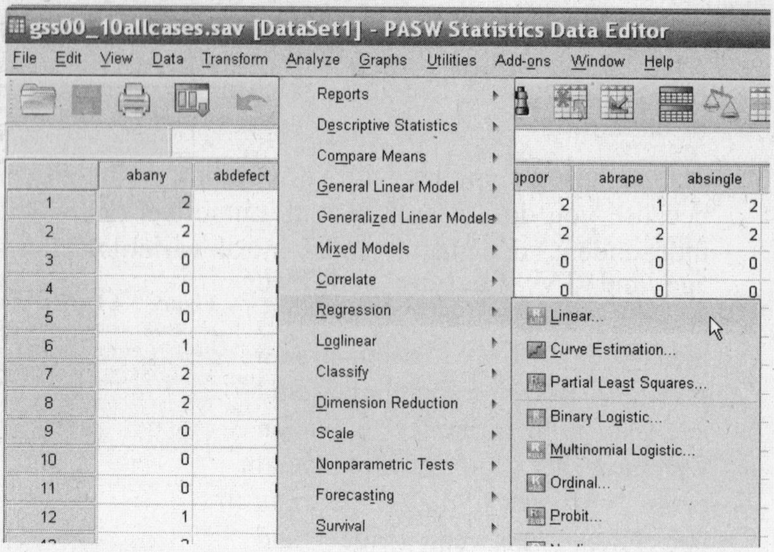

Regression analysis allows researchers to examine the impact of one or more independent variables on a given dependent variable in greater detail. *Linear regression analysis* affords a graphical picture of the relationship between two variables and produces a regression equation that can be used to predict values of the dependent variable given values of the independent variable. *Multiple regression* allows the researcher to more clearly specify the unique impact of each of several independent variables on a dependent variable. The basic regression equation is *YN* = *a* + *b*(*X*), where *a* equals the value of Y when *X* is 0, *b* represents the slope of the regression line, *X* is a given value on the independent variable, and *YN* is the estimated value on the dependent variable. In short, regression analysis is a technique for establishing the regression equation representing the geometric line that comes closest to the distribution of data points on a graph.

Figure 3.7 Linear Regression Dialog Box

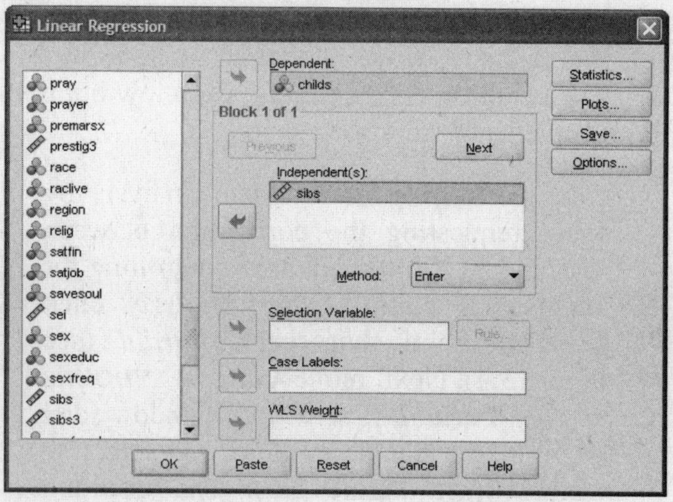

Let's do a regression analysis for the connection between the number of siblings one has and the number of children one has. Click ***Analyze***, click ***Regression***, and then click ***Linear*** (see Figure 3.6). Then move **CHILDS** to the ***Dependent*** selection window and **SIBS** to the ***Independent(s)*** window (see Figure 3.7). The results should resemble Figure 3.8. Focus on the last box, coefficients. Another term for the *a* value is *constant*. So, **CHILDS** = 1.39 + (.12 H **SIBS**). So, someone with two siblings would be predicted to have 1.63 children (1.63 = 1.39 + (.12 H 2). If you would like to look at the results graphically, called a *scatterplot*, follow these instructions: click ***Graphs***, click ***Scatter***, click ***Simple***, click ***Define***. Move **CHILDS** to the Y axis window and **SIBS** to the X axis window, and click ***OK***. Now to the really exciting stuff: multiple independent variables. Multiple regression analysis allows you to determine the effect of each of several independent variables on a dependent variable, with the remaining independent variables controlled. This is helpful in identifying which independent variables matter most for a dependent variable, while simultaneously controlling for other independent variables. Let's do an example. Simply click additional variables **AGE** and **SEI** (social class) into the ***Independent(s)*** window (if **SIBS** isn't already there, add it, and if **CHILDS** isn't in the ***Dependent*** window, add it). The last table in your output will resemble Figure 3.9. The *beta* values, also called *standardized coefficients*, are the ones to examine because they show the relative impact of an independent variable on the dependent variable with the other independent variables controlled. We see that age is the most salient for predicting the number of children someone has (.37), that the number of siblings has a small but positive impact (.14), and that social class has a negligible negative impact (-.10).

Figure 3.8 Linear Regression Output

Model Summary

Model	R	R Square	Adjusted R Square	Std. Error of the Estimate
1	.213[a]	.045	.045	1.658

a. Predictors: (Constant), sibs NUMBER OF BROTHERS AND SISTERS

ANOVA[b]

Model		Sum of Squares	df	Mean Square	F	Sig.
1	Regression	630.801	1	630.801	229.432	.000[a]
	Residual	13265.862	4825	2.749		
	Total	13896.663	4826			

a. Predictors: (Constant), sibs NUMBER OF BROTHERS AND SISTERS

b. Dependent Variable: childs NUMBER OF CHILDREN

Coefficients[a]

Model		Unstandardized Coefficients		Standardized Coefficients	t	Sig.
		B	Std. Error	Beta		
1	(Constant)	1.388	.038		36.472	.000
	sibs NUMBER OF BROTHERS AND SISTERS	.124	.008	.213	15.147	.000

a. Dependent Variable: childs NUMBER OF CHILDREN

Figure 3.9 Multiple Regression Output

Coefficients[a]

Model		Unstandardized Coefficients		Standardized Coefficients	t	Sig.
		B	Std. Error	Beta		
1	(Constant)	.251	.093		2.710	.007
	sibs NUMBER OF BROTHERS AND SISTERS	.081	.008	.141	10.155	.000
	sei RESPONDENT SOCIOECONOMIC INDEX	-.009	.001	-.099	-7.208	.000
	age AGE OF RESPONDENT	.036	.001	.372	27.331	.000

a. Dependent Variable: childs NUMBER OF CHILDREN

Chapter 4

Data Input and Modification

Now the less exciting, but still important, phase of data analysis: data input and modification. The data input and data modification stages are important because they are crucial for building an accurate and useful data set for the types of analyses described earlier. Often choices must be made on how to code and prepare the data for analysis.

Data Input

Let's begin with a sample data set of college students for you to enter into SPSS. We'll use a few items from the General Social Survey data set used in this volume and fill in some fake data. For all items, let's code a 9 (or 99 for AGE) if someone didn't respond to an item.

Recall that the dialog box in Figure 1.1 appears when you start SPSS. This time click ***Type in Data***, which opens the ***Data Editor***. On the bottom left of your screen you will see two tabs. The ***Variable View*** tab enables you to specify various aspects of your variablesBname, type, etc. The ***Data View*** tab is the place where you actually input the data. Click the ***Variable View*** tab if it is not already highlighted and let's define our variables. Notice that SPSS fills in default values for each variable.

Copy the names for the variables from the abbreviated codebook below to the fields under ***Name***. Keep variable names to eight letters and numbers when you need to construct them in the future. The ***Type*** of all our variables is numeric, so no change is needed. The maximum number of digits our variables take up is only two, but leave the ***Width*** values for all variables at the default value of 8 for now. We have no decimal values, so change the number of places after the decimal point to 0 under ***Decimals***. Fill in an appropriate label for each variable under ***Label.*** Then fill in labels for the values under ***Values*** as shown in the codebook below (skip for CASE). Do so by first clicking on the appropriate field (it will show the default ***None***), then click the small grey box at the right of that field. Fill in "1" after ***Value*** and "male" after ***Value Label*** for the SEX variable (skip the quotation marks), and then click the ***Add*** button. Follow the same procedure for value 2, "female." Click ***OK*** when done with a variable. No need to add values for AGE because the raw value is itself meaningful – it tells how old a respondent is. Add value labels for the remaining variables.

CASE ID number for each separate case.
SEX Respondent's sex; 1) male, 2) female.
AGE Respondent's age at last birthday.
LEVEL Class level; 1) first year, 2) sophomore, 3) junior, 4) senior, 8) other.
ABANY Support for abortion if the woman wants it for any reason; 1) yes, 2) no.
HAPPY Level of happiness; 1) very happy, 2) pretty happy, 3) not too happy.

CASE	SEX	AGE	LEVEL	ABANY	HAPPY
1	1	18	1	2	3
2	1	24	3	1	1
3	2	31	4	1	9
4	2	99	1	1	1
5	2	22	2	1	2
6	1	20	2	2	2
7	2	26	3	2	2
8	1	42	2	2	1
9	1	36	4	1	1
10	2	21	9	5	2

Recall that we specified a value of 9 (99 for **AGE**) if someone failed to answer an item. We fill in these values under the ***Missing*** heading. **CASE** has no missing values, so we'll leave that field unchanged from the ***None*** default. Click on the field for **SEX,** click the grey box on the right within that field, click ***Discrete Missing Values*** within the dialog box, fill in "9" in the first value box, then click ***OK.*** Complete missing values for the remaining variables (use 99 for **AGE**). Perhaps later you will work with a variable that has only a few cases in one of the values, and you may wish to simply define that value as missing to remove the cases from the analyses. For example, if you have only 3 graduate students in your study of 1,000 college students, you may wish to treat them as missing data so that your conclusions will apply only to undergraduates.

Under the ***Columns*** heading, just leave the default values of 8. The ***Align*** heading affects the appearance of the data values – just leave the default values at "right." Later you can experiment with choosing "left" or "center" alignment. The ***Measure*** heading allows us to indicate if a variable is nominal, ordinal, or "scale" (interval or ratio). Select "scale" for **CASE** and **AGE**, select "nominal" for **SEX**, **LEVEL**, and **ABANY**, and select "ordinal" for **HAPPY**. Your screen should resemble Figure 4.1.

Figure 4.1 Data Editor View for Sample Codebook

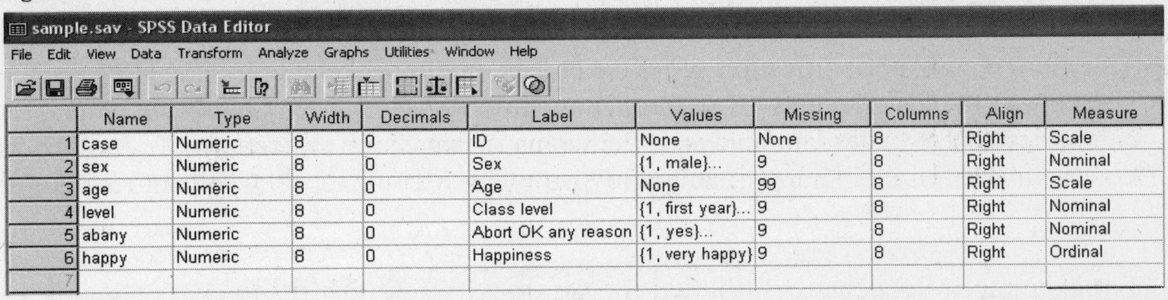

sample.sav - SPSS Data Editor

File Edit View Data Transform Analyze Graphs Utilities Window Help

	Name	Type	Width	Decimals	Label	Values	Missing	Columns	Align	Measure
1	case	Numeric	8	0	ID	None	None	8	Right	Scale
2	sex	Numeric	8	0	Sex	{1, male}...	9	8	Right	Nominal
3	age	Numeric	8	0	Age	None	99	8	Right	Scale
4	level	Numeric	8	0	Class level	{1, first year}...	9	8	Right	Nominal
5	abany	Numeric	8	0	Abort OK any reason	{1, yes}...	9	8	Right	Nominal
6	happy	Numeric	8	0	Happiness	{1, very happy}	9	8	Right	Ordinal
7										

Good work! You are now ready to input the sample fake data included in the codebook above. Click the ***Data View*** tab on the bottom left. Do one case at a time; each case is a row. So, in the first row (the first case), fill in values of 1 for CASE, 1 for SEX, 18 for AGE, 1 for LEVEL, 2 for ABANY, and 3 for HAPPY. Fill in the data for the remaining cases. Your screen should resemble Figure 4.2. If you make a mistake for a particular value, simply click on the erroneous box and enter a correction in the entry field and press ***Enter***.

Figure 4.2 Data Editor View for Sample Data Set

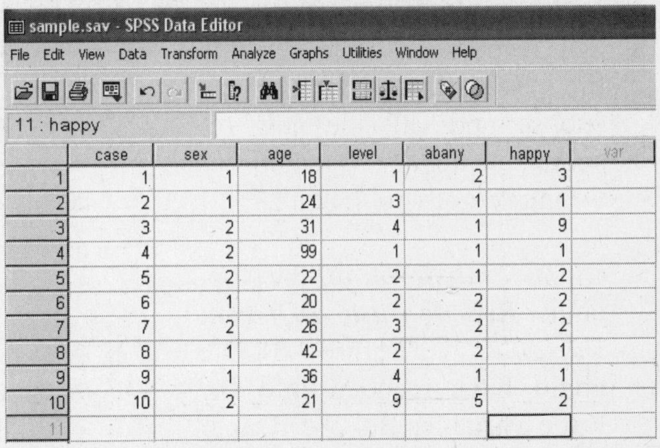

sample.sav - SPSS Data Editor

File Edit View Data Transform Analyze Graphs Utilities Window Help

11 : happy

	case	sex	age	level	abany	happy	var
1	1	1	18	1	2	3	
2	2	1	24	3	1	1	
3	3	2	31	4	1	9	
4	4	2	99	1	1	1	
5	5	2	22	2	1	2	
6	6	1	20	2	2	2	
7	7	2	26	3	2	2	
8	8	1	42	2	2	1	
9	9	1	36	4	1	1	
10	10	2	21	9	5	2	
11							

To save the data file, click on the ***Save*** icon on the top left and give your file a name, such as ***Sample***. Later on, when you have several files to look at, you can go from whatever file you are working on to a new file or an existing file by clicking the ***File*** option on the top left. There you can work with data files and output files. Upon exit, SPSS will ask you if you want to save the contents of the ***Data Editor*** if you have made any changes to the data. Do so unless you have made some temporary changes that you do not wish to save. SPSS will also ask if you want to save the contents of the ***Output Viewer***. Perhaps not if you have copied the results you needed. Perhaps so in case you want to go back and check for further details later.

Data cleaning is the first step after setting up a data set. Use the ***Frequencies*** procedure for all variables in the sample data set, except for CASE. Make sure that the missing values are appropriately labeled in the output. Look over the frequency distributions for anomalies. Oh, what's this? You've found a value of 5 for ABANY, which has only two legitimate values (1 and 2), plus a 9 for a missing value. Must be a coding error. In real life, you could use the case ID number to locate the original survey to determine the coding error and the correct value. Then you could go into ***Data View*** and type the correct value over the incorrect one.

Modifying Data

The most common modification social scientists employ is recoding data. Perhaps a few people are at the extremes of a frequency distribution and you wish to collapse the categories to facilitate data analysis. Perhaps you have only a few people in each of a dozen values for a religion variable and you wish to combine them all into an Another category.

For example, a frequency distribution for **XMARSEX** shows that relatively few people – 73 out of 3,086 valid responses – selected the "Not wrong at all" response regarding their view about extramarital sex. This may be too few respondents to make meaningful comparisons in cross-tabulations.

Figure 4.3 Recode Command

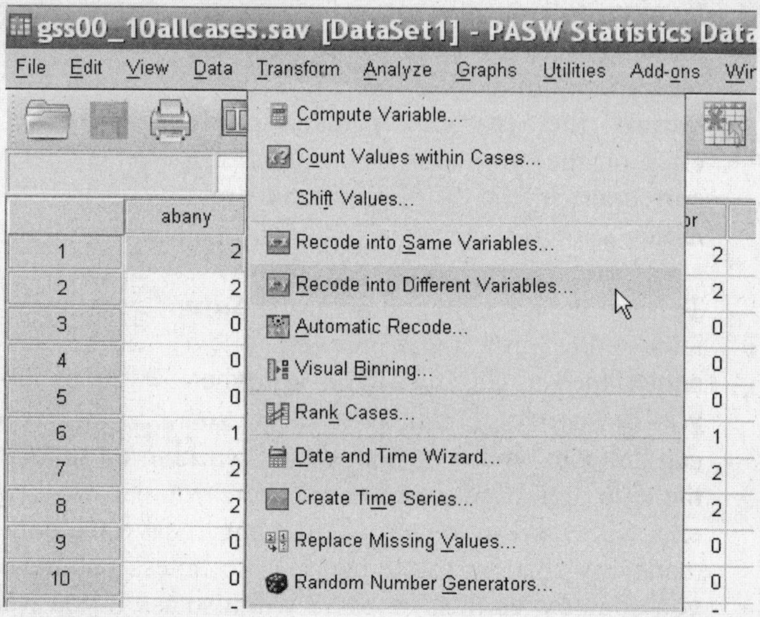

Let's combine them with the 211 respondents who selected the "sometimes wrong" response and call the new response category "sometimes wrong or not wrong at all."

Click ***Transform*** at the top, click ***Recode into Different Variables*** (see Figure 4.3). We could have recoded into the same variable. But the danger in this choice is that the original coding is gone forever, if we accidentally choose to save the data file on exit. We would be unable later to look more carefully at those who selected only the "not wrong at all" response to **XMARSEX.**

Figure 4.4 Recode Dialog Box

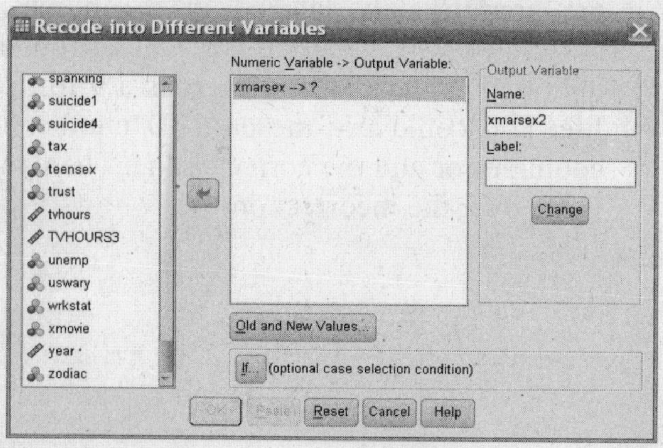

Move **XMARSEX** to the selection window in the middle. Under ***Output Variable*** on the right, fill in the new ***Name*** of **XMARSEX2** and the new ***Label*** of "extramarital sex recoded" (see Figure 4.4). Click the ***Change*** button. Next, click the ***Old and New Values*** button, fill in 1 for the ***Old*** Value and fill in 1 for the ***New Value*** and click the ***Add*** button, fill in 2 for the ***Old Value*** and fill in 2 for the ***New Value*** and click the ***Add*** button, fill in 3 for the ***Old Value*** and fill in 3 for the ***New Value*** and click the ***Add*** button, and fill in 4 for the ***Old Value*** and fill in 3 for the ***New***

Value and click the ***Add*** button (see Figure 4.5). Click ***Continue*** and click ***OK***. With some variables (like AGE), you may wish to recode a range of values into a single new value. For example, you could recode ages 18-35 into a new value of 1, which you will label "18-35" or "young." To do this, simply click ***Range*** in the ***Old and New Values*** dialog box.

Figure 4.5 Recode Old and New Values Dialog Box

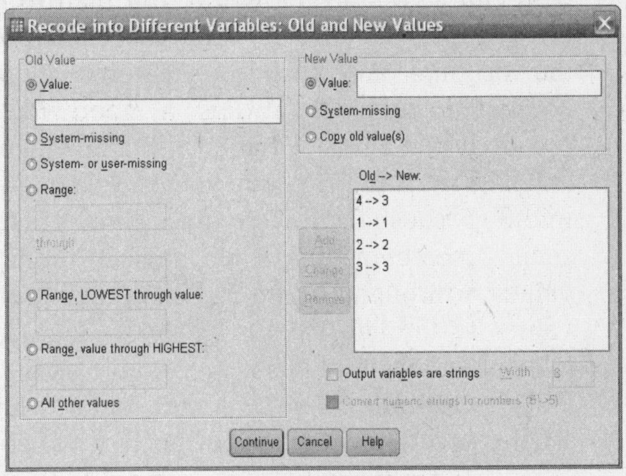

To add value labels to the new variable, click ***Variable View*** on the bottom left and follow the instructions above for adding value labels. The first two values will be identical to those in XMARSEX. Because we combined the responses to the third and fourth values in XMARSEX, make the new value label for the third response in XMARSEX2 "sometimes wrong or not wrong at all." Perform a ***Frequencies*** to make sure the recode was done correctly by making sure the total in the combined category equals the sum of the two combined categories. We can now use our new version of the extramarital variable in crosstabulations or other analyses.

Another common data modification strategy is to combine several variables into an index. For example, we could construct a social interaction index by combining SOCFREN, SOCOMMUN, and SOCREL, which measure how often respondents spend a social evening with friends outside the neighborhood, with someone in the neighborhood, and with relatives. We can simply add the scores together to create the index. Click ***Transform*** on the top and then click ***Compute***. Type the name of the new variable – let's call it INTERACT – in the ***Target Variable*** window in the top left. Move SOCFREND to the ***Numeric Expression*** window, click the + sign on the calculator, move SOCOMMUN to the ***Numeric Expression*** window, click the + sign, and finally move SOCREL to the ***Numeric Expression*** window (see Figure 4.6). Click ***OK***.

Figure 4.6 Compute Dialog Box

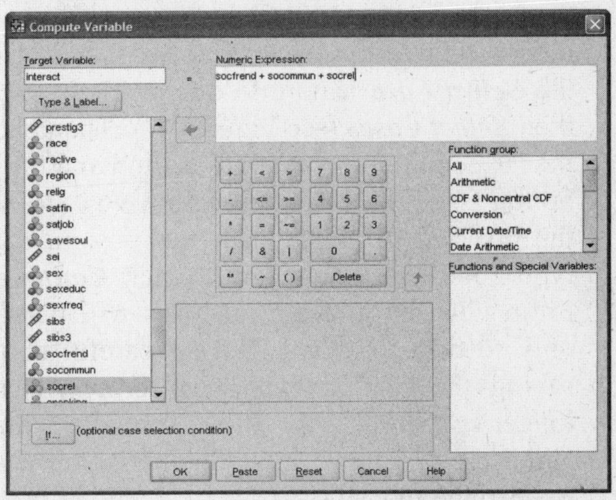

Now perform a ***Frequencies*** analysis on the new variable, INTERACT, so you can see if the calculations were done correctly. The three source variables have the same five values, so INTERACT could range from a minimum of 3 to a maximum of 15. In fact, it does. Note that the three source variables were reverse scored – a score of 1 reflects "several times a week or more" and a score of 5 reflects "once a year or less." So, our new index also is reverse scored – the lower scores reflect greater social interaction. Let's add some details to our new variable: Click ***Variable*** View on the bottom left, find INTERACT as the last variable listed, under ***Decimals*** change the 2 to 0, and under ***Label*** fill in a label (such as "Social Interaction").

You can recode INTERACT if you plan to do crosstabulations. If you want to retain both the original and recoded versions of the computed variable (always a good idea), recode it into a new variable with a new name (such as INTER3). Recode values 3 through 7 into 3 (because the lower values show greater social interaction, the 3 will reflect "high"), values 8 through 10 into 2 ("moderate"), and values 11 through 15 into 1 ("low"). You can add some details to this new variable as well: Click ***Variable View*** on the bottom left, where INTER3 will now be the last variable listed. Under ***Decimals*** change the 2 to 0, and under ***Label*** fill in a label (such as "Social Interaction in Thirds"). Under ***Values*** create values for 1 (low), 2 (moderate), and 3 (high).

Figure 4.7 Select Cases Command

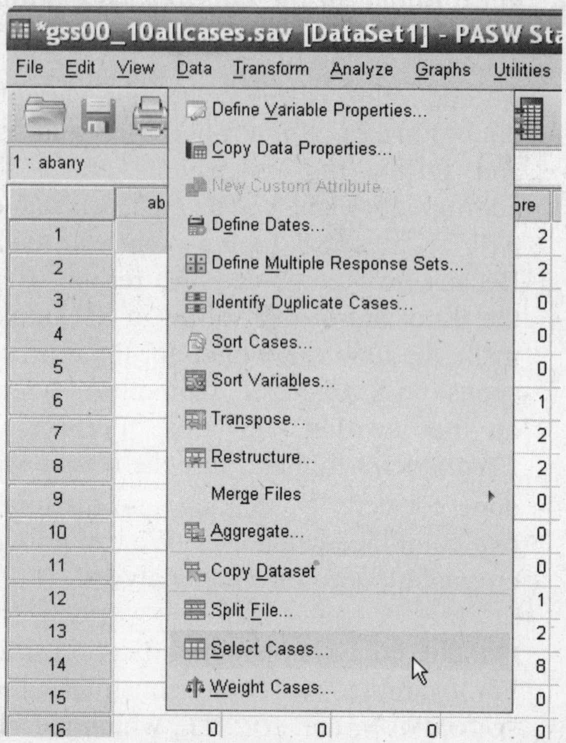

Occasionally we wish to examine a subset of a data set in greater detail. For example, the General Social Survey combines data for 2000 and 2010. A lot happened between those dates that may affect people's responses. So, we may wish to do a detailed analysis of how such factors as race, age, sex, and marital status affected something like attitudes towards abortion or gun control separately for the two years. Or perhaps we are simply interested in only one of the two years. Use the ***Select Case*** feature to do so. Click ***Data***, then ***Select Cases*** (see Figure 4.7). Then click the "If condition is satisfied" button and the ***If*** button (see Figure 4.8). Now move YEAR into the box on the right, click the = sign, then type 2000 (see Figure 4.9). Click ***Continue***. From now on, all your analyses will pertain only to data for 2000. But be careful about saving the file on exit because it will no longer contain data for 2010. To address this, simply save it under a new name if you wish to keep the file.

Figure 4.8 Select Cases Dialog Box

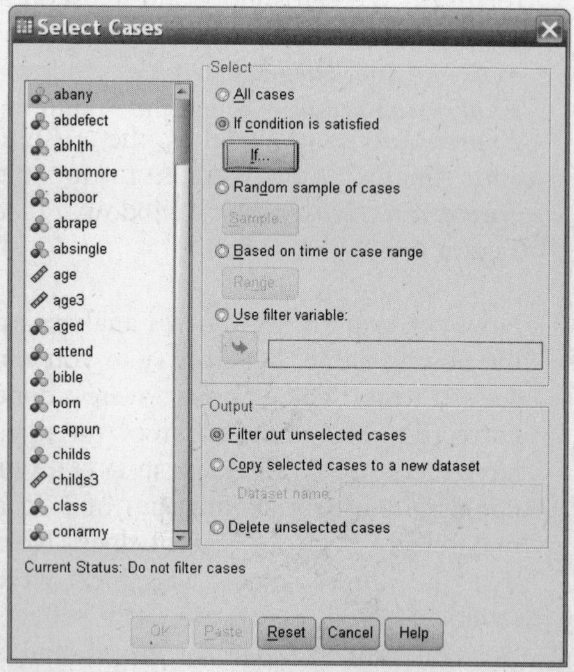

Congratulations! You are now SPSS proficient. Be sure to list this skill on your resume (it may even help you get a job!). We've covered all the basic moves in SPSS. Of course, many more options exist in SPSS for data modification and analysis. But the ones we have not covered follow the basic logic and commands that we have covered. With appropriate instruction in methods and statistics, you should be able to use them with little difficulty.

Figure 4.9 If Command in Select Cases

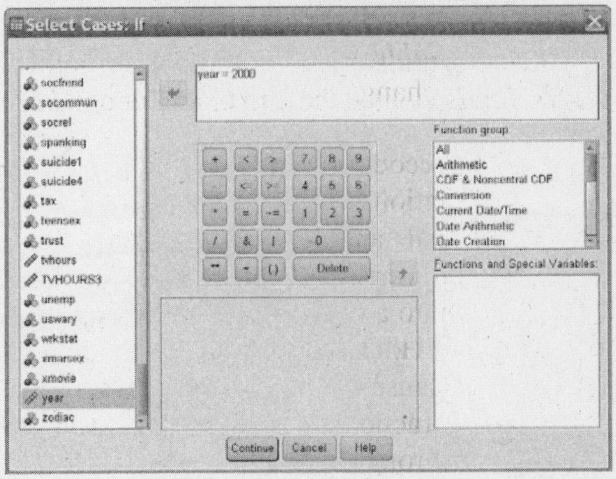

Chapter 5

Chapter Keyed Exercises

Now it's your chance to apply what you have learned. These exercises are keyed to the chapters in Earl Babbie's *The Basics of Social Research, 6th* edition.

Part 1 An Introduction to Inquiry

Chapter 1 Human Inquiry and Science

Exercise 1.1 Select three variables in the General Social Survey data set that could be considered independent variables, factors that may influence or cause something else. Explain why you consider them to be independent variables. Perform a frequency distribution on each and request a different type of chart for each. Interpret the results.

Exercise 1.2 Select three variables in the General Social Survey data set that could be considered dependent variables, factors that may be influenced or caused by the three independent variables you picked for the first exercise. Explain why you consider them to be dependent variables. Perform a frequency distribution on each and request a different type of chart for each. Interpret the results.

Exercise 1.3 Babbie talks about social regularities in this chapter. Scan the codebook for the General Social Survey (see Appendix) and identify a social regularity that you can test, which would mean connecting an independent variable with a dependent variable. Use SPSS to determine if your hypothesized social regularity holds. You may end up running a crosstab, comparing means, or running a correlation.

Chapter 2 Paradigms, Theory, and Research

Exercise 2.1 Review Babbie's analysis of the conflict paradigm. Using deductive theory construction, we could hypothesize that more successful people will be more satisfied with the way things are because they benefit more from current social arrangements. We can test this hypothesis by examining the relationship between occupational prestige and overall happiness. Run a crosstabulation of **HAPPY** by **SEI3** (which is **SEI**Boccupational prestigeBrecoded into thirds). Be sure to put **SEI3** on top (columns) and **HAPPY** on the side (rows), and request column percentages. Present your table and analyze the percentage differences to see if the hypothesis is correct.

Exercise 2.2 Use Babbie's discussion of critical race theory to test a hypothesis linking race to a relevant dependent variable. Use SPSS to test your hypothesis. You may end up running a crosstab, comparing means, or running a correlation.

Exercise 2.3 Use Babbie's discussion of the feminist paradigm to test a hypothesis linking an independent variable of your choice to one of the items pertaining to women's roles: **FECHLD, FEFAM, FEPOL,** or **FEPRESCH**. Use SPSS to test your hypothesis. You may end up running a crosstab, comparing means, or running a correlation.

Chapter 3 The Ethics and Politics of Social Research

Exercise 3.1 Review the General Social Survey description in the Appendix, including the introduction. Determine how adequately this study addresses each of the ethical issues Babbie addresses: voluntary participation, no harm to subjects, anonymity and confidentiality, the researcher's identity, and analysis and reporting.

Exercise 3.2 Assess the General Social Survey from the standpoint of Babbie's discussion about objectivity and ideology. How might the answers given by respondents be affected by some of the factors he discusses?

Part 2 The Structuring of Inquiry

Chapter 4 Research Design

Exercise 4.1 Babbie discusses the three criteria for establishing nomothetic causality. Your assignment is to apply these criteria to a crosstabulation of attitudes regarding the legalization of marijuana by gender. Run a crosstabulation of **GRASS** by **SEX**. Be sure to put **SEX** on top (columns) and **GRASS** on the side (rows), and request column percentages. Summarize the relationship as reflected in the percentage differences. Apply the three criteria for establishing causality to thc relationship between support for the legalization of marijuana and gender.

Exercise 4.2 The General Social Survey contains data for 2000 and 2010. Hence, it allows you to perform a longitudinal analysisBa trend analysis. Examine how attitudes toward abortion have changed over those time periods. Run a crosstabulation of **ABANY** (approve abortion for any reason) by **YEAR**. Be sure to put **YEAR** on top (columns) and **ABANY** on the side (rows), and request column percentages. Summarize the percentage differences across the years. Develop an explanation for the changes.

Exercise 4.3 The General Social Survey also allows for a cohort analysis because you can compare people who were younger in 2000 with other people who also were about 10 years older when surveyed in 2010. For simplicity's sake, compare people with a value of 2 on **AGE3** (37-53) in 2000 with people with a value of 3 (54-89) on **AGE 3** in 2010 on a variable of interest. Simply do a crosstab of **AGE3** with your dependent variable and add **YEAR** as a control variable. Analyze the results from a cohort perspective. Realize that respondents in the second group will not be exactly 10 years older because of the particular range of ages in the **AGE3** categories.

Chapter 5 Conceptualization, Operationalization, and Measurement

Exercise 5.1 The split-half method for establishing reliability involves splitting your items into two halves and then seeing how well the scores on those two halves correlate. This exercise involves determining how reliably three items in the General Social Survey measure attitudes toward gender roles. As a modified split-half approach, examine how three of the four indicators correlate. That is, run crosstabulations of **FECHLD** with **FEFAM**, **FECHLD** with **FEPOL**, and **FEFAM** with **FEPOL**. Because we are not suggesting that one is an independent variable and the other a dependent variable, you can put whichever ones you wish on the top (columns) and side (rows). But request both column and row percentages to facilitate your analyses. Report the three tables and analyze the degree to which these three items reliably measure attitudes towards gender roles.

Exercise 5.2 Assess how adequately the four items on gender roles (**FECHLD, FEFAM, FEPOL,** and **FEPRESCH**) measure attitudes toward gender roles. Consider both the conceptualization process and the operationalization/measurement process. Apply two of the types of validity that Babbie discusses to this measure of attitudes toward gender roles.

Exercise 5.3 Perform a modified version of criterion validity by validating the attitudinal item about gun permits against gun ownership. Presumably gun owners are less likely to support a law requiring people to get a police permit before purchasing a gun. Consider support for a law requiring permits (**GUNLAW**) to be the independent variable and place it on top (columns). Consider gun ownership (**OWNGUN**) to be the dependent variableBthe criterion variableBand place it on the side (rows). Request column percentages. Does gun ownership validate the attitudinal item on gun laws? Explain.

Exercise 5.4 The interchangeability of indicators means that if different indicators measure the same concept, they will all have a similar connection with another variable. The larger concept of confidence in institutions is measured with the following indicators: **CONARMY, CONBUS, CONEDUC, CONFED, CONLEGIS, CONPRESS, CONSCI, CONLABOR, CONCLERG,** and **CONFINAN**. Examine how people of different races (**RACE**) compare on three of these confidence items and determine if the interchangeability of indicators applies.

Chapter 6 Indexes, Scales, and Typologies

The chapter reviews four basic steps in constructing an index: (1) selecting possible items, (2) examining their empirical relationships, (3) combining some items into an index, and (4) validating the index. You are to apply these steps to three items in the General Social Survey that measure attitudes toward abortion. The items are **ABRAPE, ABHLTH,** and **ABDEFECT**.

Exercise 6.1 Regarding item selection, assess these items in terms of face validity, unidimensionality, and variance. To assess variance, you will have to do a frequency distribution on the items. Use the Avalid percent@ column for your analyses because this column excludes missing values.

Exercise 6.2 Assess the bivariate relationships among these items by doing the following crosstabulations: **ABRAPE** with **ABHLTH**, **ABRAPE** with **ABDEFECT**, and **ABHLTH** with **ABDEFECT**. Request both column and row percentages. Present the tables and interpret the results.

Exercise 6.3 Regarding the multivariate (in this case trivariate) relationships among the items, use **ABHLTH** as the criterion variable. For the sake of simplicity, the results should be presented in the following form, although other forms would also be appropriate. Remember that column percentages will not necessarily total 100% because the numbers in your table will only represent those who say Ayes@ to the **ABHLTH** item. Interpret the results.

AYes@ response on **ABHLTH**

		ABRAPE	
		Yes	No
ABDEFECT	Yes	__%	__%
	No	__%	__%

Exercise 6.4 Regarding index scoring, apply the discussion in the text to this example by creating a new variable, **ABORT**, by summing the three items. Recode the variables first so that ANo@ = 0 and AYes@ = 1. Use the ***Recode*** function to recode the three variables into new variables (you might use **ABRAP**, **ABHLT**, and **ABDEF** as names). Recode 0 into 9 (0 is Amissing,@ so you are just recoding one missing value into anotherB9Bso that the 0 can be used for the Ano@ value), 1 into 1, and 2 into 0. Hence, 1 will still reflect Ayes,@ but 0 will now reflect Ano.@ The index will reflect the number of items the respondent agreed with, among those who responded to all three items. Use the ***Compute*** function by simply adding together your recoded variables. Present a frequency distribution of your index and interpret the results.

Exercise 6.5 Regarding index validation via item analysis, validate **ABRAPE** by using the index **ABORT** (created in Exercise 6.4) as the independent (column) variable and **ABRAPE** as the dependent (row) variable. Do a crosstabulation and request column percentages. Present and analyze the results.

Exercise 6.6 Regarding index validation via external validation, validate the index with political views by running a crosstabulation of **POLVIEWS** by the index you created in Exercise 6.4, **ABORT**. Make **POLVIEWS** the row variable and **ABORT** the column variable and request column percentages. Present and analyze the results.

Exercise 6.7 Babbie discusses various alternatives for addressing missing values. One of those is to examine other responses of those who have missing values on a variable. See if race and gender differences exist for those who are missing on **DIVLAW**, the item asking if the respondent feels that divorce should be easier or more difficult to obtain. That is, run crosstabulations of **DIVLAW** by **RACE** and **SEX**. Make **DIVLAW** the row variable and **RACE** and **SEX** the column variables (run two tables), and request column percentages.

The data file already excludes the missing values from the results because these values have already been defined as missing. Similarly, the missing values are not noted in the codebook in the Appendix. There are two missing values, 8 for Adon=t know@ and 9 for Anot applicable.@ Examine only those who responded Adon=t know.@ So, first change the coding so that this value, 8, is not defined as missing before running your tables. Do this by first clicking ***Data Editor***, then ***Variable View***, then ***Missing*** for the variable **DIVLAW**, then delete the 8 from the list of missing values. Summarize the differences by race and sex, if any, for those who responded Adon=t know@ versus those who gave one of the coded responses. (When you exit SPSS, respond "no" to the query if you wish to save the data set because you do not want to save the temporary change in missing values for **DIVLAW**.)

Chapter 7 The Logic of Sampling

Exercise 7.1 Let's take several ten-percent random samples of the General Social Survey data set and see how the values on **AGE** for the samples compare to the Apopulation@ of the entire data set. First calculate a mean on **AGE** for the entire data set by clicking ***Analyze***, ***Descriptive Statistics***, and ***Descriptives***. Record the number of cases and the mean. To draw a ten-percent sample, click ***Data*** and ***Select Cases***. In the dialog box, click ***Random Sample of Cases***. Click the ***Sample*** button. Click ***Approximately*** and fill in 10. Click ***Continue*** to return to the previous dialog box. Near the bottom, click ***Filtered*** regarding unselected cases. Click ***OK.*** Now calculate the mean for **AGE** by following the procedures above. Record the number of cases and the mean for Sample 1. To get rid of the first sample and return to the full data set, click ***Data*** and ***Select Cases***, and click ***All Cases*** near the top. Click ***OK***. Now you can proceed to take the second sample and record the number of cases and the mean for **AGE**. Take a total of ten samples in this manner and record the data.

Compare the ten means for the samples with the Apopulation@ mean you began with. Arrange the sample means in order from the greatest minus difference to the greatest plus difference above and below the mean of the complete sample. Create a histogram for age by clicking ***Analyze***, ***Descriptive Statistics***, and ***Frequencies***. Click the ***Charts*** button and select ***Histograms***. Click ***With Normal Curve*** under ***Histogram***. Click ***Continue***. Select **AGE** while in the ***Frequencies*** dialog box. Click ***OK***. The spread of your sample means should roughly follow the distribution for age in the histogram. If you took a large number of ten-percent samples, the means will tend to spread out in a way that follows the normal curve. Study this phenomenon in relationship to your text discussion of sampling distributions and the calculation of standard errors. Summarize your results.

Exercise 7.2 Read the sampling discussion about the General Social Survey in the appendix and make specific connections to the content of this chapter.

Part 3 Modes of Operation

Chapter 8 Experiments

Exercise 8.1 As a contrived type of natural experiment, we may investigate if owning a gun (the Astimulus@) affects propensity toward violence. Test this hypothesis by running a crosstabulation of **POLHITOK** (it is OK for police to strike a citizen) by **OWNGUN** (own a gun or not). Make **POLHITOK** the row variable and **OWNGUN** the column variable, and request column percentages. Report the table and analyze the results to determine if the hypothesis is correct.

Exercise 8.2 Let's try a modified type of static-group comparison design. The Astimulus@ is whether African Americans live in the respondent's neighborhood (**RACLIVE**). The Aposttest@ is respondents' views on spending to improve the conditions of Blacks (**NATRACE**). Run a crosstab and determine if there is a connection between the two factors.

Chapter 9 Survey Research

Exercise 9.1 Below is part of an interview guide used in a study. Assign variable names to the items and create an SPSS data set, complete with value labels and missing data values. Make up responses for ten respondents and input that data. Run ***Frequencies*** for the items.

Case ID_____ Respondent's gender: [ ] Male [ ] Female

Hello. My name is _________ and I'm conducting a short survey as part of an exercise for my research methods class. Would you be willing to participate in the study? The interview will take less than five minutes. (ENCOURAGE IF RELUCTANT.)

I'm going to read some opinions about some contemporary social issues. As I read each one, please tell me if you agree or disagree with the statement. (READ EACH ITEM AND CHECK THE APPROPRIATE CATEGORY.)

	Agree	Disagree	Don't Know	No Answer
1. Homosexual marriages should be made legal.	[]	[]	[]	[]
2. The government should make it harder for people to obtain welfare	[]	[]	[]	[]
3. Health care for all young children should be provided at no cost to parents.	[]	[]	[]	[]

I am going to name some institutions in this country. As far as the PEOPLE RUNNING these institutions are concerned, would you say that you have a great deal of confidence, only some confidence, or hardly any confidence at all in them? (CODE FOR EACH ITEM; CODE DK FOR ADON'T KNOW@ OR NA FOR ANO ANSWER@ IF APPROPRIATE.)

	Great Deal	Only Some	Hardly Any	DK	NA
4. Local government	[]	[]	[]	[]	[]
5. Religion	[]	[]	[]	[]	[]
6. Science	[]	[]	[]	[]	[]

Exercise 9.2 Compare the question wording used in the General Social Survey with the advice given in the text. Note how the items do and do not reflect that advice.

Exercise 9.3 A few items in the General Social Survey contain volunteered responses (**COURTS, DIVLAW, AGED**). Select one and run a frequencies. Discuss the effects of not offering such a response as part of the question.

Chapter 10 Qualitative Field Research

Exercise 10.1 Explore software for compiling and analyzing qualitative data by visiting http://onlineqda.hud.ac.uk/Intro_CAQDAS/index.php. Explain why SPSS is less suitable for qualitative analysis. Also explain how SPSS could be used for some types of qualitative research.

Chapter 11 Unobtrusive Research

Exercise 11.1 Assess the General Social Survey in terms of Babbie's discussion of the issues important to consider when using existing statistics.

Exercise 11.2 Devise a content analysis study of letters to the editor in a newspaper familiar to you. Develop the coding system for tapping manifest and latent content. Construct SPSS variables names and values, and explain how and why you made the choices that you did.

Exercise 11.3 Assess the reliability and validity of the General Social Survey as a source of existing statistics, using the ideas raised in the chapter.

Chapter 12 Evaluation Research

Exercise 12.1 Explore people's thoughts about whether they think we're spending too much money on various national problems, too little money, or about the right amount. Run ***Frequencies*** on the following variables: **NATAID, NATCRIME, NATARMS, NATEDUC, NATFARE,** and **NATRACE**. Present your results. Based on the public's response to these issues, select one area that you feel most needs attention and develop an evaluation research study to examine the success of a hypothetical social policy program in that area.

Exercise 12.2 Devise a modified evaluation research study to examine if respondents' mothers worked or not when the respondents were children (**MAWRKGRW**) and how that affects respondents' attitudes about women needing to stay home and take care of the family (**FEFAM**) and if preschool children suffer when their mothers are employed (**FEPRESCH**).

Exercise 12.3 Which items in the General Social Survey could be used as social indicators? Why? Run a frequencies on one and draw conclusions about the item as a social indicator.

Part 4 Analysis of Data

Chapter 13 Qualitative Data Analysis

Exercise 13.1 (repeats *Exercise 10.1*) Explore software for compiling and analyzing qualitative data by visiting http://onlineqda.hud.ac.uk/Intro_CAQDAS/index.php. Explain why SPSS is less suitable for qualitative analysis. Also explain how SPSS could be used for some types of qualitative research.

Chapter 14 Quantitative Data Analysis

Exercise 14.1 Babbie discusses the usefulness of collapsing response categories to facilitate analyses. Apply this principle to the variable **REGION** in the General Social Survey. Run a frequency distribution on **REGION** and interpret the results. Be sure to use the Avalid percent@ category, which omits missing values. Then collapse **REGION** as follows: ANew England@ and AMiddle Atlantic@ into AEast;@ AEast North Central@ and AWest North Central@ into ANorth Central;@ ASouth Atlantic,@ AEast South Central,@ and AWest South Central@ into ASouth;@ and AMountain@ and APacific@ into AWest.@ Recalculate the percentages. Report the frequency distribution *before* collapsing and summarize the results. Be sure to use the Avalid percent@ category. Report the frequency distribution *after* collapsing and summarize the results. Compare the results for the two tables and note why the collapsed table may be more useful for some analysis purposes. (Do not save the data file on exit so that the original values will remain intact.)

Exercise 14.2 Select eight variables that interest you from the General Social Survey codebook in the Appendix. Select several items that could be used as independent variables and several that could be used as dependent variables. Produce frequency distributions of the variables. Request measures of central tendency.

Interpret the frequency distributions; use the Avalid percent@ column because it omits missing data. Interpret the appropriate statistics for five of the variables. Your interpretations should mention the level of measurement of each variable and refer only to those statistics appropriate for that level. (Note that computers will calculate any requested statistics on your data, whether appropriate or not.)

Exercise 14.3 Construct six tables from the variables you analyzed in ***Exercise 14.2.*** Select two tables to interpret. Be sure to 1) compare percentages on the dependent variable across categories of the independent variable, 2) summarize the relationship, and 3) briefly explain the relationship by providing possible reasons for why the relationship exists or does not exist.

Exercise 14.4 Select two ratio or interval level variables in the General Social Survey and use the appropriate SPSS procedure to calculate the standard deviation. Interpret one of them according to the guidelines in the text.

Chapter 15 Reading and Writing Social Research

Exercise 15.1 How would someone with expertise in SPSS read and assess a research report differently than someone without that expertise? Assume equal levels of methods competence.

Appendix

General Social Survey 2000 and 2010

The General Social Survey (GSS) is a survey done by the National Opinion Research Center on the attitudes, behaviors, and background characteristics of Americans. These surveys have occurred annually or biannually since 1972. Samples are full probability samples, and the samples reflect multistage area probability samples to the block level. At the block level, however, quota sampling is used with quota based on sex, age, and employment status. The primary sampling units employed are Standard Metropolitan Statistical Areas or nonmetropolitan counties, stratified by region, age, and race before selection. The units of selection at the second stage are block groups and enumeration districts, stratified according to race and income before selection. The third stage of selection involved blocks, which were selected with probabilities proportional to size. The final sampling units involved housing units and individuals.

The sample sizes for the 2000 and 2010 studies were 2,817 and 2,044. The "split-ballot" technique was used, where some questions were asked of part of the sample and other questions were asked of another part of the sample. Basic questions were asked of all participants. As a result, the totals for some of the variables you analyze may not total the numbers of completed cases noted above.

The data files you will be using will vary according to the version of SPSS (or other statistical program) that you will use. Some earlier "student" versions of SPSS are limited to 50 variables and 1,500 cases (student versions of SPSS are not available in recent versions of SPSS). If you will be using an earlier version of SPSS with a student version, you will be using either a combined 2000-2010 data file containing 750 cases from each year or separate data files of 1,500 cases for each year. In both situations, you will have access only to the 50 variables in the codebook with asterisks. The subset data files are random samples taken from the full data files.

You may wish to limit your analyses to only one year, do separate analyses for the two years, or combine both years in one analysis. Unless you specifically select a particular year, your analyses will include data from both years (if you are using the combined years file). Be very clear on which subgroups you will use in your analysis. If you use the combined sample of cases across both years to study attitudes toward abortion, for example, changes that occurred in attitudes between the years will not be evident and may even mask relationships you wish to examine between, say, gender and attitudes toward abortion. That is, if women were more likely to support abortion in one year and men in the other year, a table relating gender to abortion support for the combined samples would show a negligible gender difference. All the variables in the files were asked in both 2000 and 2010.

Be sure you distinguish clearly between independent and dependent variables in your analyses. Many of the typically independent variables, such as gender, age, and race, are located toward the beginning of the data file. Some variables, such as how much people watch TV or their level of education, could be either independent or dependent, depending on your research question. Select an appropriate number of what you consider independent and dependent variables, or as many of each type as indicated by your instructor.

A tip for making it easier to find variables in SPSS: change the default listing of variables to show the variable names. That way you can easily find the variables in the codebook below. Click ***Edit***, ***Options***, ***General***, and then in the top left click ***Display Names***. To make it easier to match the values with labels for those values, in ***Options*** click ***Output Labels*** and then select ***Names and Labels*** under "Variables in item labels shown as" and right under that select ***Values and Labels*** for "Variable values in item labels shown as."

Once again, an asterisk indicates variables which ***are*** available in reduced size files. ***All*** the variables are available in the regular files.

* = available in smaller files.

YEAR*

1. Year of sample.

   2000 or 2010

REGION*

2. Region of interview.
   1. New England
   2. Middle Atlantic
   3. East North Central
   4. West North Central
   5. South Atlantic
   6. East South Central
   7. West South Central
   8. Mountain
   9. Pacific

RACE*

3. What race do you consider yourself?
   1. White
   2. Black
   3. Other

SEX*

4. Sex (Coded by interviewer)
   1. Male
   2. Female

BORN

5. Were you born in this country?

   1. Yes
   2. No

MOBILE16 6. When you were 16 years old, were you living in this same (city/town/county)?

1. Same state, same city
2. Same state, different city
3. Different state

AGE* 7. Age. (Determined by asking date of birth, actual ages recorded). Values reflect actual age (except last category of 89 include 89 and older).

AGE3* 8. AGE variable recoded into thirds.

1. 18-36
2. 37-53
3. 54-89

MARITAL* 9. Are you currently married, widowed, divorced, separated, or have you never been married? (Divorced and separated combined into one category)

1. Married
2. Widowed
3. Divorced/separated
4. Never married

DEGREE* 10. Highest degree.

0. Less than high school
1. High school
2. Associate/junior college
3. Bachelor's
4. Graduate

REALINC 11. Family income in constant dollars (responses are dollars of income adjusted so you can compare across years).

REALINC3 12. Family income in constant dollars in thirds.

1. Low
2. Moderate
3. High

REALRINC 13. Respondent's income in constant dollars (responses are dollars of income adjusted so you can compare across years).

REALRIN3 14. Respondent's income in constant dollars in thirds.

1. Low
2. Moderate
3. High

FINRELA 15. Compared with American families in general, would you say your family income is far below average, average, above average, or far above average? (Recoded into three categories.)

1. Below average
2. Average
3. Above average

FINALTER* 16. During the last few years, has your financial situation been getting better, worse, or has it stayed the same?

1. Better
2. Worse
3. Stayed same

WRKSTAT 17. Last week were you working full time, part time, going to school, keeping house, or what?

1. Working full time
2. Working part time
3. With a job, but not at work because of temporary illness, vacation, strike
4. Unemployed, laid off, looking for work
5. Retired
6. In school
7. Keeping house
8. Other

JOBFIND 18. About how easy would it be for you to find a job with another employer with approximately the same income and fringe benefits you now have? Would you say very easy, somewhat easy, or not easy at all?

1. Very easy
2. Somewhat easy
3. Not easy at all

JOBLOSE 19. Thinking about the next 12 months, how likely do you think it is that you will lose your job or be laid off – very likely, fairly likely, not too likely, or not at all likely?

1. Very likely
2. Fairly likely
3. Not too likely
4. Not at all likely

DWELOWN 20. Do you/does your family own your (home/apartment), pay rent, or what? ("Other" recoded to missing.)

1. Own or is buying
2. Rent

SATJOB 21. On the whole, how satisfied are you with the work that you do – would you say that you are very satisfied, moderately satisfied, a little dissatisfied, or very dissatisfied? (Responses recoded into three categories.)

1. Very satisfied
2. Moderately satisfied
3. Dissatisfied

SATFIN* 22. We are interested in how people are getting along financially these days. So far as you and your family are concerned, would you say that you are pretty well satisfied with your present financial situation, more or less satisfied, or not satisfied at all?

1. Pretty well satisfied
2. More or less satisfied
3. Not satisfied at all

GOODLIFE* 23. The way things are in America, people like me and my family have a good chance of improving our standard of living – do you agree or disagree?

1. Strongly agree
2. Agree
3. Neither agree nor disagree
4. Disagree
5. Strongly disagree

KIDSSOL* 24. When your children are at the age you are now, do you think their standard of living will be much better, somewhat better, about the same, somewhat worse, or much worse than yours is now?

1. Much better
2. Somewhat better
3. About the same
4. Somewhat worse
5. Much worse
6. No children (volunteered)

PARSOL 25. Compared to your parents when they were the age you are now, do you think your own standard of living now is much better, somewhat better, about the same, somewhat worse, or much worse than theirs was?

1. Much better
2. Somewhat better
3. About the same
4. Somewhat worse
5. Much worse

UNEMP 26. At any time during the last ten years, have you been unemployed and looking for work for as long as a month?

1. Yes
2. No

SEI* 27. Hodge/Siegel/Rossi prestige scale score for respondent's occupation. (Actual score recorded)

SEI3* 28. SEI, respondent's occupational prestige, recoded into thirds.
1. Low
2. Moderate
3. High

CLASS 29. If you were asked to name one of four names for your social class, which would you say you belong in: the lower class, the working class, the middle class, or the upper class?
1. Lower class
2. Working class
3. Middle class
4. Upper class

GRANBORN 30. Were all of your four grandparents born in this country? (Actual number born outside the United States recoded to "one through three" and "all four.")
0. All born in the United States
1. One through three born outside the United States
2. All four born outside the United States

RELIG* 31. What is your religious preference? It is Protestant, Catholic, Jewish, some other religion, or no religion?
1. Protestant
2. Catholic
3. Jewish
4. None
5. Other

ATTEND* 32. How often do you attend religious services? (Recoded into four categories.)
1. Less than once a year
2. Once a year through several times a year
3. About once a month through almost weekly
4. Every week or more

PRAY 33. About how often do you pray?
1. Several times a day
2. Once a day
3. Several times a week
4. Once a week
5. Less than once a week
6. Never

BIBLE 34. Which of these statements comes closest to describing your feelings about the Bible?

1. The Bible is the actual word of God and is to be taken literally, word for word
2. The Bible is the inspired word of God but not everything in it should be taken literally, word for word
3. The Bible is an ancient book of fables, legends, history, and moral precepts recorded by men

POSTLIFE 35. Do you believe there is a life after death?

1. Yes
2. No

CHILDS* 36. How many children have you ever had? Please count all that were born alive at any time (including any you had from a previous marriage).

0-7. Actual number (e.g., 3 means 3 children)
8. Eight or more

CHILDS3 37. CHILDS, number of children, recoded into thirds.

1. None
2. One or two
3. Three or more

SIBS* 38. How many brothers and sisters did you have? Please count those born alive, but no longer living, as well as those alive now. Also include stepbrothers and stepsisters, and children adopted by your parents. Values reflect actual number of siblings.

SIBS3 39. SIBS, number of siblings, recoded into thirds.

1. 0-2
2. 3-4
3. 5 or more

PARTYID* 40. Generally speaking, do you usually think of yourself as a Republican, Democrat, Independent, or what? (Recoded into four categories.)

1. Democrat
2. Independent
3. Republican
4. Other party

POLVIEWS* 41. I'm going to show a seven-point scale on which the political views that people might hold are arranged from extremely liberal – point 1 – to extremely conservative – point 7. Where would you place yourself on this scale? (Recoded into three categories.)

1. Liberal
2. Moderate
3. Conservative

ZODIAC 42. Astrological sign of respondent.

1. Aries
2. Taurus
3. Gemini
4. Cancer
5. Leo
6. Virgo
7. Libra
8. Scorpio
9. Sagittarius
10. Capricorn
11. Aquarius
12. Pisces

TVHOURS* 43. On the average day, about how many hours do you personally watch television? (Actual numbers of hours recorded.)

TVHOURS3 44. **TVHOURS**, number of hours watching television per day, recoded into thirds.

1. 0 through 1 hours
2. 2 through 3 hours
3. 4 through 24 hours

NEWS 45. How often do your read the newspaper – every day, a few times a week, once a week, less than once a week, or never?

1. Every day
2. A few times a week
3. Once a week
4. Less than once a week
5. Never

OWNGUN* 46. Do you happen to have in your home any guns or revolvers?

1. Yes
2. No

MAWRKGRW 47. Did your mother ever work for pay for as long as a year, while you were growing up?

1. Yes
2. No

HEALTH* 48. Would you say your health, in general, is excellent, good, fair, or poor?

1. Excellent
2. Good
3. Fair
4. Poor

Please tell me whether or not *you* think it should be possible for a pregnant woman to obtain a *legal* abortion if. . .(applies to items 49-55)

ABANY* 49. The woman wants it for any reason?

1. Yes
2. No

ABDEFECT* 50. There is a strong chance of serious defect in the baby?
1. Yes
2. No

ABHLTH* 51. The woman's own health is seriously endangered by the pregnancy?
1. Yes
2. No

ABNOMORE 52. She is married and does not want any more children?
1. Yes
2. No

ABPOOR 53. The family has a very low income and cannot afford any more children?
1. Yes
2. No

ABRAPE* 54. She became pregnant as a result of rape?
1. Yes
2. No

ABSINGLE* 55. She is not married and does not want to marry the man?
1. Yes
2. No

AGED* 56. As you know, many older people share a home with their grown children. Do you think this is generally a good idea or a bad idea?
1. A good idea
2. Bad idea
3. Depends (volunteered)

PILLOK 57. Do you strongly agree, agree, disagree, or strongly disagree that methods of birth control should be available to teenagers between the ages of 14 and 16 if their parents do not approve?
1. Strongly agree
2. Agree
3. Disagree
4. Strongly disagree

I am going to name some institutions in this country. As far as the *people running* these institutions are concerned, would you say you have a great deal of confidence, only some confidence, or hardly any confidence at all in them? (Applies to items 58-67.)

CONARMY 58. Military?
1. A great deal
2. Only some
3. Hardly any

CONBUS 59. Major companies?
1. A great deal
2. Only some
3. Hardly any

CONEDUC* 60. Education?
1. A great deal
2. Only some
3. Hardly any

CONFED 61. Executive branch of the federal government?
1. A great deal
2. Only some
3. Hardly any

CONLEGIS 62. Congress?
1. A great deal
2. Only some
3. Hardly any

CONPRESS* 63. Press?
1. A great deal
2. Only some
3. Hardly any

CONSCI 64. Scientific community?
1. A great deal
2. Only some
3. Hardly any

CONLABOR 65. Organized labor?
1. A great deal
2. Only some
3. Hardly any

CONCLERG 66. Organized religion?
1. A great deal
2. Only some
3. Hardly any

CONFINAN 67. Banks and financial institutions?

1. A great deal
2. Only some
3. Hardly any

GRASS* 68. Do you think the use of marijuana should be made legal or not?

1. Should be made legal
2. Should not be made legal

CAPPUN* 69. Do you favor or oppose the death penalty for persons convicted of murder?

1. Favor
2. Oppose

DIVLAW* 70. Should divorce in this country be easier or more difficult to obtain than it is now?

1. Easier
2. More difficult
3. Stay as it (volunteered)

GUNLAW* 71. Would you favor or oppose a law which would require a person to obtain a police permit before he or she could buy a gun?

1. Favor
2. Oppose

SOCFREND* 72. How often do you spend a social evening with friends who live outside the neighborhood? (Collapsed from 7 categories.)

1. Several times a week or more
2. Several times a month
3. Once a month
4. Several times a year
5. Once a year or less

SOCOMMUN* 73. How often do you spend a social evening with someone who lives in your neighborhood? (Collapsed from 7 categories.)

1. Several times a week or more
2. Several times a month
3. Once a month
4. Several times a year
5. Once a year or less

SOCREL 74. How often do you spend a social evening with relatives? (Collapsed from 7 categories.)

1. Several times a week or more
2. Several times a month
3. Once a month
4. Several times a year
5. Once a year or less

HAPPY* 75. Taken all together, how would you say things are these days – would you say that you are very happy, pretty happy, or not too happy?
1. Very happy
2. Pretty happy
3. Not too happy

HAPMAR 76. Taking things all together, how would you describe your marriage? Would you say that your marriage is very happy, pretty happy, or not too happy?
1. Very happy
2. Pretty happy
3. Not too happy

LIFE 77. In general, do you find life pretty exciting, routine, or dull?
1. Exciting
2. Routine
3. Dull

SEXEDUC 78. Would you be for or against sex education in the public schools?
1. For
2. Against

TRUST 79. Generally speaking, would you say that most people can be trusted or that you can't be too careful in life?
1. Most people can be trusted
2. Can't be too careful
3. It depends (volunteered)

HELPFUL* 80. Would you say that most of the time people try to be helpful, or that they are mostly just looking out for themselves?
1. Try to be helpful
2. Just look out for themselves
3. It depends (volunteered)

FAIR 81. Do you think most people would try to take advantage of you if they got a chance, or would they try to be fair? ("Depends" response deleted.)
1. Would take advantage of you
2. Would try to be fair

GETAHEAD 82. Some people say that people get ahead by their own hard work; others say that luck breaks or help from other people are more important. Which do you think is most important?
1. Hard work
2. Hard work and luck equally important
3. Luck or help from others

HARMGOOD 83. Overall, modern science does more harm than good.
1. Agree
2. Neither agree nor disagree
3. Disagree

POLHITOK 84. Are there any situations you can imagine in which you would approve of a policeman striking an adult male citizen?
1. Yes
2. No

FEAR 85. Is there any area right around here–that is, within a mile–where you would be afraid to walk alone at night?
1. Yes
2. No

PREMARSX* 86. There's been a lot of discussion about the way morals and attitudes about sex are changing in this country. If a man and a woman have sex relations before marriage, do you think it is always wrong, almost always wrong, wrong only sometimes, or not wrong at all?
1. Always wrong
2. Almost always wrong
3. Wrong only sometimes
4. Not wrong at all

TEENSEX 87. What if they are in their early teens, say 14 to 16 years old? In that case, do you think sex relations before marriage are always wrong, almost always wrong, wrong only sometimes, or not wrong at all?
1. Always wrong
2. Almost always wrong
3. Wrong only sometimes
4. Not wrong at all

HOMOSEX* 88. What about sexual relations between two adults of the same sex–do you think it is always wrong, almost always wrong, wrong only sometimes, or not wrong at all?
1. Always wrong
2. Almost always wrong
3. Wrong only sometimes
4. Not wrong at all

XMARSEX 89. What is your opinion about a married person having sexual relations with someone other than the marriage partner–is it always wrong, almost always wrong, wrong only sometimes, or not wrong at all?
1. Always wrong
2. Almost always wrong
3. Wrong only sometimes
4. Not wrong at all

SEXFREQ* 90. About how often did you have sex in the last twelve months?

0. Not at all
1. Once a year through once a month
2. Two or three times a month
3. Weekly
4. Two or three times a week or more

XMOVIE 91. Have you seen an X-rated movie in the last year?

1. Yes
2. No

We are faced with many problems in this country, none of which can be solved easily or inexpensively. I'm going to name some of these problems, and for each one I'd like you to tell me whether you think we're spending too much money on it, too little money, or about the right amount. Are we spending too much, too little, or about the right amount on...(Applies to items 92-99.)

NATAID 92. Foreign aid?

1. Too little
2. About right
3. Too much

NATCRIME* 93. Halting the rising crime rate?

1. Too little
2. About right
3. Too much

NATARMS 94. The military, armaments, and defense?

1. Too little
2. About right
3. Too much

NATEDUC* 95. Improving the nation's education system?

1. Too little
2. About right
3. Too much

NATCITY 96. Solving the problems of the big cities?

1. Too little
2. About right
3. Too much

NATDRUG 97. Dealing with drug addiction?

1. Too little
2. About right
3. Too much

NATFARE 98. Welfare?

1. Too little
2. About right
3. Too much

NATRACE 99. Improving the condition of Blacks?

1. Too little
2. About right
3. Too much

FECHLD* 100. Please tell me whether you strongly agree, agree, disagree, or strongly disagree with this statement: a working mother can establish just as warm and secure a relationship with her children as a mother who does not work.

1. Strongly agree
2. Agree
3. Disagree
4. Strongly disagree

FEFAM* 101. Please tell me whether you strongly agree, agree, disagree, or strongly disagree with this statement: it is much better for everyone involved if the man is the achiever outside the home and the woman takes care of the home and family.

1. Strongly agree
2. Agree
3. Disagree
4. Strongly disagree

FEPRESCH 102. Please tell me whether you strongly agree, agree, disagree, or strongly disagree with this statement: a preschool child is likely to suffer if his or her mother works.

1. Strongly agree
2. Agree
3. Disagree
4. Strongly disagree

FEPOL* 103. Tell me if you agree or disagree with this statement: most men are better suited emotionally for politics than are most women.

1. Agree
2. Disagree

HELPPOOR 104. Some people think that the government in Washington should do everything possible to improve the standard of living of all poor Americans. Other people think it is not the government's responsibility, and that each person should take care of himself. (Five-point scale collapsed to three categories).

1. Government should improve living standards
2. Agree with both answers
3. People should take care of themselves

HELPSICK 105. In general, some people think that it is the responsibility of the government in Washington to see to it that people have help in paying for doctors and hospital bills. Others think that these matters are not the responsibility of the federal government and that people should take care of these things themselves. (Five-point scale collapsed into three categories).

1. Government should help
2. Agree with both answers
3. People should take care of themselves

SUICIDE1* 106. Do you think that a person has the right to end his or her life if this person has an incurable disease?

1. Yes
2. No

SUICIDE4 107. Do you think that a person has the right to end his or her life if this person is tired of living and ready to die?

1. Yes
2. No

LETDIE1* 108. When a person has a disease that cannot be cured, do you think doctors should be allowed by law to end the patient's life by some painless means if the patient and his family request it?

1. Yes
2. No

SPANKING 109. Do you strongly agree, agree, disagree, or strongly disagree that it is sometimes necessary to discipline a child with a good, hard spanking?

1. Strongly agree
2. Agree
3. Disagree
4. Strongly disagree

USWARY 110. Do you expect the United States to fight in another world war within the next ten years?

1. Yes
2. No

PORNLAW 111. Which of these statements comes closest to your feelings about pornography laws?

1. There should be laws against the distribution of pornography whatever the age.
2. There should be laws against the distribution of pornography to persons under 18.
3. There should be no laws forbidding the distribution of pornography.

PRAYER 112. The United States Supreme Court has ruled that no state or local government may require the reading of the Lord's Prayer or Bible verses in public schools. What are your views on this–do you approve or disapprove of the court ruling?

1. Approve
2. Disapprove

RACLIVE 113. Are there any Blacks/African-Americans living in this neighborhood now?

1. Yes
2. No

TAX 114. Do you consider the amount of federal income tax which you have to pay as too high, about right, or too low?

1. Too high
2. About right
3. Too low

COURTS 115. In general, do you think the courts in this area deal too harshly or not harshly enough with criminals?

1. Too harshly
2. Not harshly enough
3. About right (volunteered)

EQWLTH* 116. Some people think that the government in Washington ought to reduce the income differences between the rich and the poor, perhaps by raising the taxes of wealthy families or by giving income assistance to the poor. Others think that the government should not concern itself with reducing this income difference between the rich and the poor.

Here is a card with a scale from 1 to 7. Think of a score of 1 as meaning that the government ought to reduce the income differences between rich and the government ought to reduce the income differences between rich and poor, and a score of 7 meaning that the government should not concern itself with reducing income differences. What score between 1 and 7 comes closest to the way you feel? (CIRCLE ONE):

1. Government should do something to reduce the income differences between the rich and poor
2.
3.
4.
5.
6.
7. Government should not concern itself with income differences